On-Chip Networks

Second Edition

Synthesis Lectures on Computer Architecture

Editor
Margaret Martonosi, *Princeton University*

Founding Editor Emeritus
Mark D. Hill, *University of Wisconsin, Madison*

Synthesis Lectures on Computer Architecture publishes 50- to 100-page publications on topics pertaining to the science and art of designing, analyzing, selecting and interconnecting hardware components to create computers that meet functional, performance and cost goals. The scope will largely follow the purview of premier computer architecture conferences, such as ISCA, HPCA, MICRO, and ASPLOS.

On-Chip Networks, Second Edition
Natalie Enright Jerger, Tushar Krishna, and Li-Shiuan Peh
2017

Space-Time Computing with Temporal Neural Networks
James E. Smith
2017

Hardware and Software Support for Virtualization
Edouard Bugnion, Jason Nieh, and Dan Tsafrir
2017

Datacenter Design and Management: A Computer Architect's Perspective
Benjamin C. Lee
2016

A Primer on Compression in the Memory Hierarchy
Somayeh Sardashti, Angelos Arelakis, Per Stenström, and David A. Wood
2015

Research Infrastructures for Hardware Accelerators
Yakun Sophia Shao and David Brooks
2015

On-Chip Networks, Second Edition

Natalie Enright Jerger, Tushar Krishna, and Li-Shiuan Peh

ISBN: 978-3-031-00627-2   paperback
ISBN: 978-3-031-01755-1   ebook

DOI 10.1007/978-3-031-01755-1

A Publication in the Springer series
SYNTHESIS LECTURES ON ADVANCES IN AUTOMOTIVE TECHNOLOGY

Lecture #40
Series Editor: Margaret Martonosi, *Princeton University*
Founding Editor Emeritus: Mark D. Hill, *University of Wisconsin, Madison*
Series ISSN
Print 1935-3235   Electronic 1935-3243

On-Chip Networks

Second Edition

Natalie Enright Jerger
University of Toronto

Tushar Krishna
Georgia Institute of Technology

Li-Shiuan Peh
National University of Singapore

SYNTHESIS LECTURES ON COMPUTER ARCHITECTURE #40

ABSTRACT

This book targets engineers and researchers familiar with basic computer architecture concepts who are interested in learning about on-chip networks. This work is designed to be a short synthesis of the most critical concepts in on-chip network design. It is a resource for both understanding on-chip network basics and for providing an overview of state of-the-art research in on-chip networks. We believe that an overview that teaches both fundamental concepts and highlights state-of-the-art designs will be of great value to both graduate students and industry engineers. While not an exhaustive text, we hope to illuminate fundamental concepts for the reader as well as identify trends and gaps in on-chip network research.

With the rapid advances in this field, we felt it was timely to update and review the state of the art in this second edition. We introduce two new chapters at the end of the book. We have updated the latest research of the past years throughout the book and also expanded our coverage of fundamental concepts to include several research ideas that have now made their way into products and, in our opinion, should be textbook concepts that all on-chip network practitioners should know. For example, these fundamental concepts include message passing, multicast routing, and bubble flow control schemes.

KEYWORDS

interconnection networks, topology, routing, flow control, deadlock, computer architecture, multiprocessor system on chip

To our families
for their encouragement and patience
through the writing of this book.

Contents

Preface

This book targets engineers and researchers familiar with many basic computer architecture concepts who are interested in learning about on-chip networks. This work is designed to be a short synthesis of the most critical concepts in on-chip network design. We envision this book as a resource for both understanding on-chip network basics and for providing an overview of state-of-the-art research in on-chip networks. We believe that an overview that teaches both fundamental concepts and highlights state-of-the-art designs will be of great value to both graduate students and industry engineers. While not an exhaustive text, we hope to illuminate fundamental concepts for the reader as well as identify trends and gaps in on-chip network research.

With the rapid advances in this field, we felt it was timely to update and review the state of the art in this second edition. We introduce two new chapters at the end of the book, as will be detailed below. Throughout the book, in addition to updating the latest research in the past years, we also expanded our coverage of fundamental concepts to include several research ideas that have now made their way into products and, in our opinion, should be textbook concepts that all on-chip network practitioners should know. For example, these fundamental concepts include message passing, multicast routing, and bubble flow control schemes.

The structure of this book is as follows. Chapter 1 introduces on-chip networks in the context of multi-core architectures and discusses their evolution from simple point-to-point wires and buses for scalability.

Chapter 2 explains how networks fit into the overall system architecture of multi-core designs. Specifically, we examine the set of requirements imposed by cache-coherence protocols in shared memory chip multiprocessors, and contrast that with the requirements in message-passing multi-cores. In addition to examining the system requirements, this chapter also describes the interface between the system and the network.

Once a context for the use of on-chip networks has been provided through a discussion of system architecture, the details of the network are explored. As topology is often a first choice in designing a network, Chapter 3 describes various topology trade-offs for cost and performance. Given a network topology, a routing algorithm must be implemented to determine the path(s) messages travel to be delivered throughout the network fabric; routing algorithms are explained in Chapter 4. Chapter 5 deals with the flow control mechanisms employed in the network; flow control specifies how network resources, namely buffers and links, are allocated to packets as they travel from source to destination. Topology, routing, and flow control all factor into the microarchitecture of the network routers. Details on various microarchitectural trade-offs and design issues are presented in Chapter 6. This chapter includes the design of buffers, switches, and allocators that comprise the router microarchitecture. Although power consumption can be addressed through innovations in all areas of on-chip networks, we focus our new power discussion in the microarchitecture chapter as this is where many such optimizations are realized.

New Chapter 7 covers the nuts and bolts of modeling and evaluating on-chip networks, from software simulations to RTL design and emulation on FPGA, to architectural models of delay, throughput, area, and power. The chapter also guides the reader on useful metrics for evaluating on-chip networks and ideal theoretical yardsticks for comparing against.

With the plethora of industry and academia on-chip network chips now available, we dedicate a new Chapter 8 to a survey of these. The chapter provides the reader with a sweeping understanding of how the various fundamental concepts presented in the earlier chapters come together, and the implications of the design and implementation of such concepts.

Finally in Chapter 9, we leave the reader with thoughts on key challenges and new areas of exploration that will drive on-chip network research in the years to come. Substantial new research has clearly surfaced, and here we focus on various significant trends that highlight the cross-cutting nature of on-chip network research. Emerging new interconnects and devices substantially change the implementation tradeoffs of on-chip networks, and in turn prompt new designs. Newly important metrics such as resilience, due to increasing variability in the fabrication process, or quality-of-service that is prompted by multiple workloads running simultaneously on many-cores, will add new dimensions and prompt new research ideas across the community.

Natalie Enright Jerger, Tushar Krishna, and Li-Shiuan Peh
May 2017

Acknowledgments

We would like to thank Margaret Martonosi for her feedback and encouragement to create the second edition of this book. We continue to be grateful to Mark Hill for his feedback and support in crafting the previous edition. Additionally, we would like to thank Michael Morgan for the opportunity to contribute once again to this lecture series. Many thanks to Timothy Pinkston and Lizhong Chen for their detailed comments that were invaluable in improving this manuscript. Thanks to Mario Badr, Wenbo Dai, Shehab Elsayed, Karthik Ganesan, Parisa Khadem Hamedani, and Joshua San Miguel of the University of Toronto for proofreading our early drafts. Thanks to Georgia Tech students Hyoukjun Kwon, Ananda Samajdar for feedback on early drafts, and Monodeep Kar for help with literature surveys. Thanks also to the many students and instructors who have used the first edition over the years and provided feedback that lead to this latest edition.

Natalie Enright Jerger, Tushar Krishna, and Li-Shiuan Peh
May 2017

CHAPTER 1

Introduction

Since the introduction of research into multi-core chips in the late 1990s [40, 271, 336], on-chip networks have emerged as an important and growing field of research. As core counts increase, and multi-core processors emerge in diverse domains ranging from high-end servers to smartphones and even Internet of Things (IoT) gateways, there is a corresponding increase in bandwidth demand to facilitate high core utilization and a critical need for scalable on-chip interconnection fabrics. This diversity of application platforms has led to research in on-chip networks spanning a variety of disciplines from computer architecture to computer-aided design, embedded systems, VLSI, and more. Here, we provide a synthesis of critical concepts in on-chip networks to quickly bootstrap students and designers into this exciting field.

1.1 THE ADVENT OF THE MULTI-CORE ERA

The combined pressures from ever-increasing power consumption and the diminishing returns in performance of uniprocessor architectures have led to the advent of multi-core chips. With a growing number of transistors available at each new technology generation, coupled with a reduction in design complexity enabled by the modular design of multi-core chips, this multi-core wave looks set to stay. Recent years have seen every industry chip vendor releasing multi-core products with increasing core counts. This multi-core wave may lead to hundreds and even thousands of cores integrated on a single chip. We have already seen multi-core products targeting HPC with more than 50 cores on-die, and research prototypes with more than 100 cores. Heterogeneity is now common place in many market segments, in terms of the types of components that are integrated on-chip, which further ups the complexity of the on-chip interconnection fabric. Increasingly, besides processor cores, the on-chip fabric has to interconnect embedded memories, accelerators such as DSP modules, video processors, and graphics processors.

1.1.1 COMMUNICATION DEMANDS OF MULTI-CORE ARCHITECTURES

As the number of on-chip cores increases, a scalable low-latency and high-bandwidth communication fabric to connect them becomes critically important [43, 44, 95, 290]. Up to four or eight cores, buses and crossbars are the dominant interconnect. Buses are shared multi-bit physical channels that every core connects to and listens to, while one core can transmit at a time. Buses provide low-latency but poor bandwidth. Crossbars, which are described in more detail later in Chapter 6, are switches providing non-blocking connectivity between any pair of cores. They have high-bandwidth and fairly low delays, but scale poorly in terms of area and power. As a result, on-chip networks are fast replacing buses and crossbars to emerge as the pervasive communication fabric in many-core chips. Such on-chip networks have routers at every node, connected to neighbors via short local on-chip links; multiple communication flows are multiplexed over these links to provide scalability and high bandwidth. This evolution of interconnection networks as core count increases is clearly illustrated in the choice of a flat crossbar interconnect connecting all eight cores in the Sun Niagara (2005) [199], four packet-switched rings in the 9-core IBM Cell (2005) [173], and five packet-switched meshes in the 64-core Tilera TILE64 (2007) [356].

Multi-core and many-core architectures are now commonplace in a variety of computing domains. These architectures will enable increased levels of server consolidation in data centers [25, 116, 241]. Desktop applications, particularly graphics are already leveraging the multi-core wave [227, 312]. High-bandwidth communication will be required for these throughput-oriented applications. Communication latency can have a significant impact on the performance of multi-threaded workloads; synchronization between threads will require low-overhead communication in order to scale to a large number of cores. In multiprocessor systems-on-chip (MPSoCs), leveraging an on-chip network can help enable design isolation: MPSoCs utilize heterogeneous IP blocks[1] from a variety of vendors; with standard interfaces, these blocks can communicate through an on-chip network in a plug-and-play fashion.

1.2   ON-CHIP VS. OFF-CHIP NETWORKS

While on-chip networks can leverage ideas from prior multi-chassis interconnection networks[2] used in supercomputers [12, 127, 132, 311], clusters of workstations [31], and Internet routers [84], the design requirements facing on-chip networks differ starkly in magnitude; hence, novel designs are critically needed. Fortunately, by moving on-chip, the I/O bottlenecks that faced prior multi-chassis interconnection networks are alleviated substantially: the abundant on-chip wiring supplies bandwidth that is orders of magnitude higher than off-chip I/Os while obviating the inherent delay overheads associated with off-chip I/O transmission.

On the other hand, a number of stringent technology constraints present challenges for on-chip network designs. Specifically, on-chip networks targeting high-performance multi-core processors must supply high bandwidth at ultra-low latencies, with a tight power envelope and area budget. With multi-core and many-core chips, caches and interconnects compete with the cores for the same chip real estate. Integrating a large number of components under tight area and power constraints poses a significant challenge for architects to create a balance between these components.

Innovations in on-chip networks have led to communication latency that is competitive with crossbars leading to widespread adoption. Furthermore, although on-chip networks require

---

[1]IP blocks are intellectual property in the form of soft macros of reusable logic.
[2]Also referred to as off-chip interconnection networks.

much less power than buses and crossbars, they need to be carefully designed as on-chip network power consumption can be high [43, 352]. For example, up to $\sim 30\%$ of chip power is consumed by Intel's 80-core TeraFLOPS network [167, 344] and 36% by the RAW on-chip network [335]. Therefore, it is essential that power constraints be considered. For example, innovations in the Intel Single-chip Cloud Computer [160] lead to power consumption representing only 10% of total chip power. Despite improvements, power continues to be a significant concern moving the field forward. This new edition features a more in-depth discussion of power in Chapter 6.

1.3   NETWORK BASICS: A QUICK PRIMER

In the next few sections, we lay a foundation for terminology and topics covered within this book. Subsequent chapters will explore many of these areas in more depth as well as state-of-the-art research for different components of on-chip network design. Many fundamental concepts are applicable to off-chip networks as well with different sets of design trade-offs and opportunities for innovation in each domain.

Several acronyms have emerged as on-chip network research has gained momentum. Some examples are NoC (network-on-chip), OCIN (on-chip interconnection network) and OCN (on-chip networks), with NoC emerging as the pervasive acronym.

NoC
OCIN
OCN

1.3.1   EVOLUTION TO ON-CHIP NETWORKS

An on-chip network, as a subset of a broader class of interconnection networks, can be viewed as a programmable system that facilitates the transporting of data between nodes.[3] An on-chip network can be viewed as a system because it integrates many components including channels, buffers, switches and control.

With a small number of nodes, dedicated ad hoc wiring can be used to interconnect them. However, the use of dedicated wires is problematic as we increase the number of components on-chip: the amount of wiring required to directly connect every component will become prohibitive.

Designs with low core counts can leverage buses and crossbars. In both traditional multi-processor systems and newer multi-core architectures, bus-based systems scale to only a modest number of processors. This limited scalability is because bus traffic quickly reaches saturation as more cores are added to the bus, so it is hard to attain high bandwidth. The power required to drive a long bus with many cores tapping onto it is also exorbitant. In addition, a centralized arbiter adds arbitration latency as core counts increase. To address these problems, sophisticated bus designs incorporate segmentation, distributed arbitration, split transactions, and increasingly resemble switched on-chip networks.

Crossbars address the bandwidth problem of buses, and have been used for on-chip interconnects for a small number of nodes. However, crossbars scale poorly for a large number

---

[3]A node is any component that connects to the network, e.g., core, cache, memory controller, etc.

of cores; they require a large area footprint and consuming high power. For instance, the Sun Niagara 2's flat 8 × 9 crossbar interconnecting all cores and the memory controller has an area footprint close to that of a core. In response, hierarchical crossbars, where cores are clustered into nodes and several levels of smaller crossbars provide the interconnection, are used. For the 16 cores in Sun's Rock architecture, if the same flat crossbar architecture is used, it will require a 17 × 17 crossbar that will take up at least 8× more area than the final hierarchical crossbar design chosen: a 5 × 5 crossbar connecting clusters of four cores each [340]. These sophisticated crossbars resemble multi-hop on-chip networks where each hop comprises small crossbars.

On-chip networks are an attractive alternative to buses and crossbars for several reasons. First and foremost, networks represent a scalable solution to on-chip communication, due to their ability to supply scalable bandwidth at low area and power overheads that correlate sub-linearly with the number of nodes. Second, on-chip networks are very efficient in their use of wiring, multiplexing different communication flows on the same links allowing for high bandwidth. Finally, on-chip networks with regular topologies have local, short interconnects that are fixed in length and can be optimized and built modularly using regular repetitive structures, easing the burden of verification.

1.3.2   ON-CHIP NETWORK BUILDING BLOCKS

The design of an on-chip network can be broken down into its various building blocks: its topology, routing, flow control, router microarchitecture and design, and link architecture. The rest of this book is organized along these building blocks and we will briefly explain each in turn here.

Topology. An on-chip network is composed of channels and router nodes. The network topology determines the physical layout and connections between nodes and channels in the network.

Routing. For a given topology, the routing algorithm determines the path through the network that a message will take to reach its destination. A routing algorithm's ability to balance traffic (or load) has a direct impact on the throughput and performance of the network.

Flow control. Flow control determines how resources are allocated to messages as they travel through the network. The flow control mechanism is responsible for allocating (and de-allocating) buffers and channel bandwidth to waiting packets. In contrast to off-chip networks based on Ethernet technology, most on-chip networks are considered to be lossless by design.

Router microarchitecture. A generic router microarchitecture is comprised of the following components: input buffers, router state, routing logic, allocators, and a crossbar (or switch). Router functionality is often pipelined to improve throughput. Delay through each router in the on-chip network is the primary contributor to communication latency. As a result, significant research effort has been spent reducing router pipeline stages and improving throughput.

Link architecture. Most on-chip network prototypes use conventional full-swing logic and repeated wires. Full-swing wires transition from 0 V (ground) to the supply voltage when transmitting a 1, and back to ground when transmitting a 0. Repeaters (inverters or buffers) at equal intervals on a long wire are an effective technique to reduce delay, enabling delay to scale linearly with number of repeaters instead of quadratically with length.

1.3.3   PERFORMANCE AND COST

As we discuss different on-chip design points and relevant research, it is important to consider both the performance and the cost of the network. Performance is generally measured in terms of network latency or accepted traffic. For back-of-the-envelope performance calculations, zero-load latency is often used, i.e., the latency experienced by a packet when there are no other packets in the network. Zero-load latency provides a lower bound on average message latency. Zero-load latency is found by taking the average distance (given in terms of network hops) a message will travel times the latency to traverse a single hop.

    In addition to providing ultra-low latency communication, networks must also deliver high throughput. Therefore, performance is also measured by its throughput. A high saturation throughput indicates that the network can accept a large amount of traffic before all packets experience very high latencies, sustaining higher bandwidth. Figure 1.1 presents a latency vs. throughput curve for an on-chip network illustrating the zero-load latency and saturation throughput.

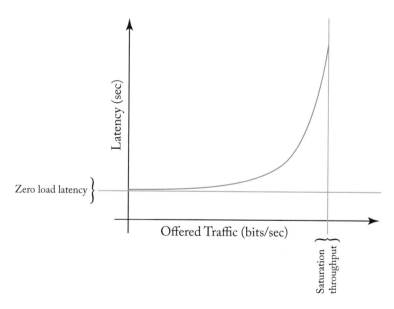

Figure 1.1: Latency vs. throughput for an on-chip network.

    The two primary costs associated with an on-chip network are area and power. As mentioned, many-core architectures operate under very tight power budgets. The impact of different designs on power and area will be discussed throughout this book, delved in more detail in Chapter 6 on Router Microarchitecture.

1.4   THIS BOOK—SECOND EDITION

This book targets engineers and researchers familiar with many basic computer architecture concepts who are interested in learning about on-chip networks. This work is designed to be a short synthesis of the most critical concepts in on-chip network design. We envision this book as a resource for both understanding on-chip network basics and for providing an overview of state-of-the-art research in on-chip networks. We believe that an overview that teaches both fundamental concepts and highlights state-of-the-art designs will be of great value to both graduate students and industry engineers. While not an exhaustive text, we hope to illuminate fundamental concepts for the reader as well as identify trends and gaps in on-chip network research.

With the rapid advances in this field, we felt it was timely to update and review the state-of-the-art in this second edition. We introduce two new chapters at the end of the book, as will be detailed below. Throughout the book, in addition to updating the latest research in the past years, we also expanded our coverage of fundamental concepts to include several research ideas that have now made their way into products and in our opinion, should be textbook concepts that all on-chip network practitioners should know. For example, these fundamental concepts include message passing, multicast routing and bubble flow control schemes.

The structure of this book is as follows. Chapter 2 explains how networks fit into the overall system architecture of multi-core designs. Specifically, we examine the set of requirements imposed by cache-coherence protocols in shared memory chip multiprocessors, and contrast that with the requirements in message-passing multi-cores. In addition to examining the system requirements, this chapter also describes the interface between the system and the network.

Once a context for the use of on-chip networks has been provided through a discussion of system architecture, the details of the network are explored. As topology is often a first choice in designing a network, Chapter 3 describes various topology trade-offs for cost and performance. Given a network topology, a routing algorithm must be implemented to determine the path(s) messages travel to be delivered throughout the network fabric; routing algorithms are explained in Chapter 4. Chapter 5 deals with the flow control mechanisms employed in the network; flow control specifies how network resources, namely buffers and links, are allocated to packets as they travel from source to destination. Topology, routing, and flow control all factor into the microarchitecture of the network routers. Details on various microarchitectural trade-offs and design issues are presented in Chapter 6. This chapter includes the design of buffers, switches and allocators that comprise the router microarchitecture. Although power consumption can be addressed through innovations in all areas of on-chip networks, we focus our new power discussion in the microarchitecture chapter as this is where many such optimizations are realized.

New Chapter 7 covers the nuts and bolts of modeling and evaluating on-chip networks, from software simulations to RTL design and emulation on FPGA, to architectural models of delay, throughput, area, and power. The chapter also guides the reader on useful metrics for evaluating on-chip networks and ideal theoretical yardsticks for comparing against.

With the plethora of industry and academia on-chip network chips now available, we dedicate a new Chapter 8 to a survey of these. The chapter provides the reader with a sweeping understanding of how the various fundamental concepts presented in the earlier chapters come together, and the implications of the design and implementation of such concepts.

Finally in Chapter 9, we leave the reader with thoughts on key challenges and new areas of exploration that will drive on-chip network research in the years to come. Substantial new research has clearly surfaced, and here we focus on various significant trends that highlight the cross-cutting nature of on-chip network research. Emerging new interconnects and devices substantially change the implementation tradeoffs of on-chip networks, and in turn prompt new designs. Newly important metrics such as resilience, due to increasing variability in the fabrication process, or quality-of-service that is prompted by multiple workloads running simultaneously on many-cores, will add new dimensions and prompt new research ideas across the community.

CHAPTER 2

Interface with System Architecture

Over the course of the last 15 years, single-processor-core computer chips have given way to multi-core chips. These multi-core and many-core systems have become the primary building blocks of computer systems, marking a major shift in the way we design and engineer these systems.

Achieving future performance gains will rely on removing the communication bottleneck between the processors and the memory components that feed these bandwidth-hungry many-core designs. Increasingly, efficient communication between execution units or cores will become a key factor in improving the performance of many-core chips.

In this chapter, we explore three major types of computer systems where on-chip networks form a critical backbone: shared-memory chip multiprocessors (CMPs) in high end servers and embedded products, message passing systems and multiprocessor SoCs (MPSoCs) in the mobile consumer market. A brief overview of general architectures and their respective communication requirements is presented.

CMP: chip multiprocessor

2.1 SHARED MEMORY NETWORKS IN CHIP MULTIPROCESSORS

Parallel programming is extremely difficult but has become increasingly important [30]. With the emergence of many-core architectures, parallel hardware is now pervasive across a range of commodity systems. The growing prevalence of parallel systems requires an increasing number of parallel applications. Maintaining a globally shared address space alleviates some of the burden placed on the programmers to write high-performance parallel code. This is because it is easier to reason about a global address space than it is for a partitioned one. A partitioned global address space (PGAS) is common in modern SMP designs where the upper address bits choose which socket the memory address is associated with.

SMP: symmetric multiprocessor

In contrast, the message passing paradigm explicitly moves data between nodes and address spaces, so programmers have to explicitly manage communications. A hybrid approach that utilizes message passing (e.g., MPI) between different shared-memory nodes with a partitioned address space is common in massively parallel processing architectures. We focus the bulk of our discussion on shared-memory CMPs since they are widely expected to be the mainstream multi-core architecture in the next few years. Like SMPs, CMPs typically have a shared global address space; however unlike SMPs, CMPs may exhibit non-uniform memory access latencies. Exceptions to the use of a shared-memory paradigm do exist. For example, the IBM Cell processor uses explicit messaging passing for DMA into local memory. The Intel SCC eschews on-chip cache coherence in favor of a message passing architecture. We will provide a brief overview of message passing systems in Section 2.2.

With the shared-memory model, communication occurs implicitly through the loading and storing of data and the accessing of instructions. As a result, the shared-memory model is an intuitive way to realize this sharing. Logically, all processors access the same shared memory, allowing each to see the most up-to-date data. Practically speaking, memory hierarchies use caches to improve the performance of shared memory systems. These cache hierarchies reduce the latency to access data but complicate the logical, unified view of memory held in the shared memory paradigm. As a result, cache coherence protocols are designed to maintain a coherent view of memory for all processors in the presence of multiple cached copies of data. Caches are designed to be transparent to the programmer. They improve performance by keeping frequently accessed data close to the processor but the programmer bears no responsibilities for managing them. This transparency is also desirable in multiprocessor systems; however, the presence of multiple caches can lead to correctness problems if different versions of the same address can reside in multiple locations at once. Cache coherence protocols are designed to solve this challenge without burdening the programmer. Cache coherence protocols maintain a single-writer, multiple-reader invariant. Cache coherence protocols manage access to shared data such that only one processor can write a cache line at one time. Multiple processors can simultaneously read a cache line without any problem. A full discussion of cache coherence and its many different flavors is outside the scope of this lecture. Interested readers are referred to Sorin et al. [325]. Therefore, it is the cache coherence protocol that governs what communication is necessary in a shared memory multiprocessor.

Figure 2.1 depicts a typical shared memory multiprocessor consisting of 64 nodes. A node contains a processor, private level 1 instruction and data caches and a second level cache that may be private or shared. Beyond the second level of cache, a third level may be incorporated on chip. This third level of cache is most commonly shared by all processors on the chip. The processor to network interface (discussed later in this chapter) and the router serve as the gateway between the local tile and other on-chip components.

Two key characteristics of a shared memory multiprocessor shape its demands on the interconnect: the cache coherence protocol that makes sure nodes receive the correct up-to-date copy of a cache line, and the cache hierarchy.

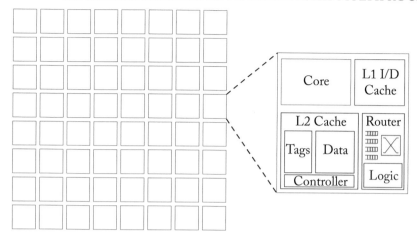

Figure 2.1: Shared memory chip multiprocessor architecture.

2.1.1 IMPACT OF COHERENCE PROTOCOL ON NETWORK PERFORMANCE

Cache coherence protocols typically enforce a single-writer, multiple-reader invariant. Any number of nodes may cache a copy of memory to read from; if a node wishes to write to that memory address, it must ensure that no other nodes are caching that address. The resulting communication requirements for a shared memory multiprocessor consist of data requests, data responses and coherence permissions. Coherence permission needs to be obtained before a node can read or write to a cache block. Depending on the cache coherence protocol, other nodes may be required to respond to a permission request.

Multiprocessor systems generally rely on one of two different types of coherence protocols: broadcast or directory, as shown in Figure 2.2. Each type of protocol results in different network traffic characteristics. Here we focus on the basics of these coherence protocols, discussing how they impact network requirements. For a more in-depth discussion of coherence, we refer the readers to other texts [85, 152, 325].

With a broadcast protocol, coherence requests are sent to all nodes on chip resulting in high bandwidth requirements. Data responses are of a point-to-point nature and do not require any ordering; broadcast systems can rely on two physical networks: one interconnect for ordering and a higher bandwidth, unordered interconnect for data transfers. Alternatively, multiple virtual channels can be used to ensure ordering among coherence traffic; requests and responses can flow through independent virtual channels [166, 200]. Figure 2.2a shows a read request resulting in a cache miss that is (1) sent to an ordering point, (2) broadcast to all cores, and then (3) receives data.

An alternative to a broadcast protocol is a directory protocol. Directory protocols do not

broadcast protocol

directory protocol

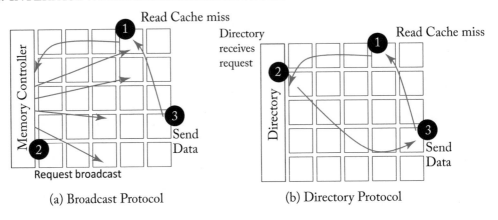

(a) Broadcast Protocol   (b) Directory Protocol

Figure 2.2: Coherence protocol network request examples.

rely on any implicit network ordering and can be mapped to an arbitrary topology. Directory protocols rely on point-to-point messages rather than broadcasts; this reduction in coherence messages allows this class of protocols to provide greater scalability. Rather than broadcast to all cores, the directory contains information about which cores have the cache block. A single core receives the read request from the directory in Figure 2.2b resulting in lower bandwidth requirements.

Directories maintain information about the current sharers of a cache line in the system and coherence state information. By maintaining a sharing list, directory protocols eliminate the need to broadcast invalidation requests to the entire system. Addresses are interleaved across directory nodes; each address is assigned a *home node*, which is responsible for ordering and handling all coherence requests to that address. Directory coherence state is maintained in memory; to make directories suitable for on-chip many-core architectures, directory caches are used. Going off-chip to memory for all coherence requests is impractical. By maintaining recently accessed directory information in on-chip directory caches, latency is reduced.

2.1.2   COHERENCE PROTOCOL REQUIREMENTS FOR THE ON-CHIP NETWORK

Cache coherence protocols require several types of messages: unicast, multicast and broadcast. Unicast (one-to-one) traffic is from a single source to a single destination (e.g., from a L2 cache to a memory controller). Multicast (one-to-many) traffic is from a single source to multiple destinations on chip (e.g., cache line invalidation messages from the directory home node to several sharers). Lastly, broadcast traffic (one-to-all) sends a message from a single source to all network destinations on chip.

With a directory protocol, the majority of requests will be unicast (or point-to-point). As a result, this places lower bandwidth demands on the network. Directory-based protocols are often chosen in scalable designs due to the point-to-point nature of communication; however, they are not immune to one-to-many or multicast communication. Directory protocols send out multiple invalidations from a single directory to nodes sharing a cache block.

Broadcast protocols place higher bandwidth demands on the interconnect as all coherence requests are of a one-to-all nature. Broadcast protocols may be required to collect acknowledgment messages from all nodes to ensure proper ordering of requests. Data response messages are point-to-point (unicast) and do not require ordering.

Cache coherent shared memory chip multiprocessors generally require two message sizes. The first message size is for coherence requests and responses without data. These messages consist of a memory address and a coherence command (request or response) and are small in size. In a cache coherence protocol, data is transferred in full cache line chunks. A data message consists of the entire cache block (typically 64 bytes) and the memory address. Both message types also contain additional network-specific data, which will be discussed in subsequent chapters.

2.1.3 PROTOCOL-LEVEL NETWORK DEADLOCK

In addition to the message types and sizes, shared memory systems require that the network be free from protocol-level deadlock. Figure 2.3 illustrates the potential for protocol level deadlock. If the network becomes flooded with requests that cannot be consumed until the network interface initiates a reply, a cyclic dependence can occur. In this example, if both processors generate a burst of requests that fill the network resources, both processors will be stalled waiting for remote replies before they can consume additional outstanding requests. If replies utilize the same network resources as requests, those replies cannot make forward progress resulting in deadlock.

protocol-level deadlock

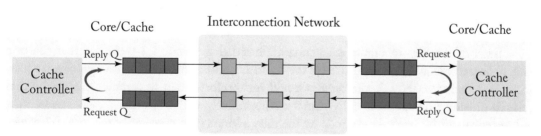

Figure 2.3: Protocol-level deadlock. Figure adapted from [294].

Protocols can require several different message classes. Each class contains a group of coherence actions that are independent of each other; that is, a request message in one class will not lead to the generation of another request message in the same class, but can trigger a message of a different class. Deadlock can occur when there are resource dependences between messages of different classes [322]. Here we describe three typical classes: requests, interventions, and

message classes

responses. Request messages include loads, stores, upgrades, and writebacks. Interventions are messages sent from the directory to request modified data be transferred to a new node. Examples of response messages include invalidation acknowledgments, negative acknowledgments (indicating a request has failed) and data messages.

Multiple virtual channels can be used to prevent protocol-level deadlock. The Alpha 21364 [254] allocates one virtual channel per message class to prevent protocol-level deadlock. By requiring different message classes to use different virtual channels, the cyclic dependence between requests and responses is broken in the network. Virtual channels and techniques to deal with protocol-level deadlock and network deadlock are discussed in Chapter 5.

2.1.4   IMPACT OF CACHE HIERARCHY IMPLEMENTATION ON NETWORK PERFORMANCE

Node design can have a significant impact on the bandwidth requirements for the on-chip network. In this section, we examine the impact of the cache hierarchy and look at how many different entities will share the injection/ejection port to the network. These entities can include multiple levels of cache, directory coherence caches, and memory controllers. Furthermore, how these entities are distributed throughout the chip can have a significant impact on overall network performance.

Caches are employed to reduce the memory latency of requests. They also serve as filters for the traffic that needs to be sent into the interconnect. For the purpose of this discussion, we assume a two-level cache hierarchy. Level 1 (L1) caches are split into instruction and data cache and the level 2 (L2) cache is the last level cache and is unified. The trade-offs discussed here could be extrapolated to cache hierarchies incorporating more levels of caching. Current chip multiprocessor research employs either private L2 caches, shared L2 caches, or a hybrid private/shared cache mechanism.

Each of the tiles in Figure 2.1 can contain either a private L2 cache for that tile or a bank of shared cache. With a private L2 cache, an L1 miss is first sent to that processor's local private L2 cache; at the L2, the request could hit, be forwarded to a remote L2 cache that holds its directory, or access off-chip memory. Alternatively, with a shared L2 cache, an L1 miss will be sent to an L2 bank determined by the miss address (not necessarily the local L2 bank), where it could hit in the L2 bank or miss and be sent off-chip to access main memory. Private caches reduce the latency of L2 cache hits on chip and keep frequently accessed data close to the processor. A drawback to private caches is the replication of shared data in several caches on chip. This replication causes on-chip storage to be used less effectively. With each core having a small private L2 cache, interconnect traffic between caches will be reduced, as only L2 cache misses go to the network; however, interconnect traffic bound off chip is likely to increase (as data that do not fit in the private L2 cache will have to be evicted off chip). With a private L2 cache, the on-chip network will interface with just the L2 cache at each node as shown in

Figure 2.4a; the injection and ejection ports of the router are only connected to one component within the tile and do not need to be shared.

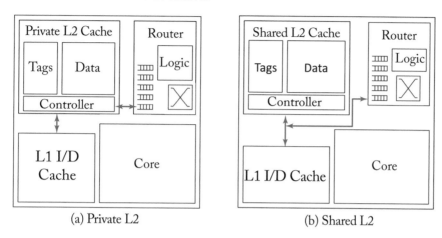

(a) Private L2   (b) Shared L2

Figure 2.4: Private and shared caches.

Figure 2.5 provides two walk-through examples of a many-core design configured with private L2 caches. In Figure 2.5a, the load of A misses in L1, but hits in the core's private L2 cache, and after step 3, the data are returned to the L1 and the core. However, in Figure 2.5b, the load of A misses in the private L2 cache and must be sent to the network interface (4), sent through the network to the memory controller (5), sent off chip, and finally re-traverse the network back to the requestor (6). After step 6, the data are installed in the L2 and forwarded to the L1 and the core. In this scenario, a miss to a private L2 cache requires two network traversals and an off-chip memory access.

Alternatively, the L2 cache can be shared amongst all or some of the cores. Shared caches represent a more effective use of storage as there is no replication of cache lines. However, L2 cache hits incur additional latency to request data from a different tile. Shared caches place more pressure on the interconnection network as L1 misses also go into the network, but more effective use of storage may reduce pressure on the off-chip bandwidth to memory. With shared caches, more requests will travel to remote nodes for data. As shown in Figure 2.4b, the on-chip network must attach to both the L1s and the L2 when the L2 is shared; both levels of cache share the injection and ejection bandwidth of the router.

Figure 2.6 provides two walk-through examples similar to those in Figure 2.5 but with a many-core system configured with a shared L2 cache. In Figure 2.6a, the L1 cache experiences a miss to address A. Address A maps to a remote bank of the shared L2, so the load request must be sent to the network interface (3) and traverse the network to the appropriate node. The read request arrives at the remote node (4) and is serviced by the L2 bank (5). The data are sent to the network interface (6) and re-traverse the network back to the requestor (7). After step 7,

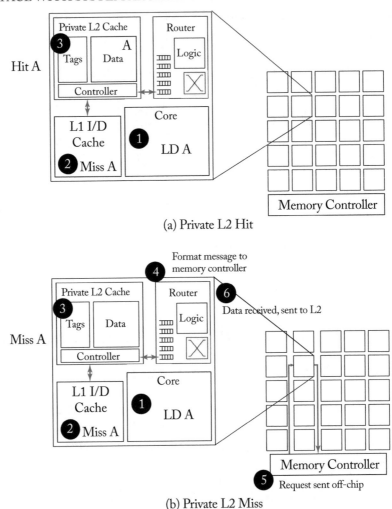

(a) Private L2 Hit

(b) Private L2 Miss

Figure 2.5: Private L2 caches walk-through example.

the data are installed in the local L1 and sent to the core. Here, an L2 hit requires two network traversals when the address maps to a remote cache (e.g., addresses can be mapped by a function $A \bmod N$, where N is the number of L2 banks).

In Figure 2.6b, we give a walk-through example for an L2 miss in a shared configuration. Initially, steps 1-4 are the same as the previous example. However, now the shared L2 bank misses to address A (5). The read request must again be sent to the network interface (6), forwarded through the network to the memory controller and sent off chip (7), returned through the network to the shared L2 bank and installed in the L2 (8) and then sent through the network

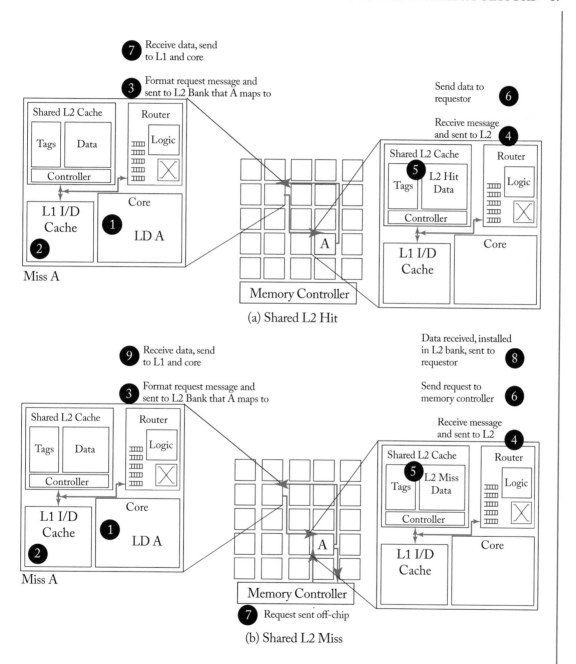

Figure 2.6: Shared L2 cache walk-through example.

back to the requestor (9). Once the data have been received (9), it can be installed in the private L1 and sent to the core. This shared miss scenario requires four network traversals to satisfy the read request.

2.1.5   HOME NODE AND MEMORY CONTROLLER DESIGN ISSUES

directory
home node

With a directory protocol, each address statically maps to a *home node*. The directory information resides at the home node which is responsible for ordering requests to all addresses that map to this *home node*. The directory either supplies the data from off chip, either from memory or from another socket, or sends intervention messages to other nodes on chip to acquire data and/or permissions for the coherence request. For a shared L2 cache, the *home node* with directory information is the cache bank that the address maps to. From the example in Figure 2.6, the directory is located at the tile marked A for address *A*. If remote L1 cache copies need to be invalidated (for a write request to *A*), the directory will send those requests through the network. With a private L2 cache configuration, there does not need to be a one-to-one correspondence between the number of *home nodes* and the number of tiles. Every tile can house a portion of the directory (*n home nodes*), there can be a single centralized directory, or there can be a number of *home nodes* in between 1 and *n*. Broadcast protocols such as the Opteron protocol [81] require an ordering point similar to a home node from which to initiate the broadcast request.

In a two-level cache hierarchy, L2 cache misses must be satisfied by main memory. These requests travel through the on-chip network to the memory controllers. Memory-intensive workloads can place heavy demands on the memory controllers making memory controllers hot-spots for network traffic. Memory controllers can be co-located with processor and cache tiles. With such an arrangement, the memory controllers will share a network injection/ejection port with the cache(s), as depicted in Figure 2.7a. Policies to arbitrate between the memory controller and the local cache for injection bandwidth are needed in this design. Alternatively, memory controllers can be placed as individual nodes on the interconnection network; with this design, memory controllers do not have to share injection/ejection bandwidth to/from the network with cache traffic (shown in Figure 2.7b). Traffic is more isolated in this scenario; the memory controller has access to the full amount of injection bandwidth. In current designs, memory controllers are often placed on the perimeter of the chip to allow close access to I/O pads.

2.1.6   MISS AND TRANSACTION STATUS HOLDING REGISTERS

miss status
handling register
(MSHR)

A processor-to-network interface is responsible for formatting network messages to handle cache misses (due to a load or store), cache line permission upgrades, and cache line evictions. Figure 2.8 depicts a possible organization for the processor-to-network interface. When a cache miss occurs, a miss status handling register (MSHR) is allocated and initialized. For example, on a read request, the MSHR is initialized to a read pending state and the message format block will create a network message. The message is formatted to contain the destination address (in

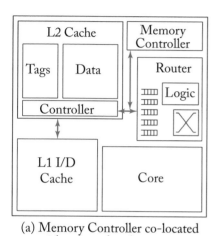

(a) Memory Controller co-located
with core and caches

(b) Memory Controller connected
to own router

Figure 2.7: Memory controllers.

the case of a directory protocol, this will be the location of the home node as determined by the memory address), the address of the cache line requested and the message request type (e.g., Read). Below the message format and send block, we show several possible message formats that may be generated depending on the type of request. When a reply message comes from the network, the MSHR matches the reply to one of the outstanding requests and completes the cache miss actions. The message receive block is also responsible for receiving request messages from the directory or another processor tile to initiate cache-to-cache transfers; the protocol finite state machine takes proper actions and formats a reply message to send back into the network. Messages received from the network may also have several different formats that must be properly handled by the message receive block and the protocol finite state machine.

The memory-to-network interface (shown in Figure 2.9) is responsible for receiving memory request messages from processors (caches) and initiating replies. Different types and sizes of messages are received from the network and sent back into the network as shown above the message format and send block and the message receive block. At the memory side, transaction status handling registers (TSHRs) handle outstanding memory requests. If memory controllers are guaranteed to service requests in order, the TSHRs could be simplified to a FIFO queue.

transaction status
handling register
(TSHR)

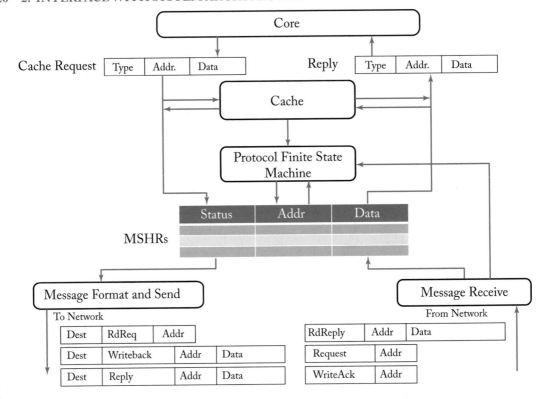

Figure 2.8: Processor-to-network interface (adapted from Dally and Towles [86]).

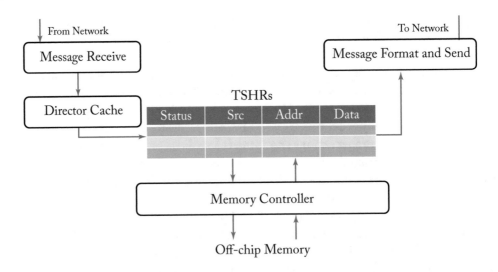

Figure 2.9: Memory-to-network interface (adapted from Dally and Towles [86]).

However, as memory controllers often reorder memory requests to improve utilization, a more complicated interface is required. Once a memory request has been completed, the message format and send block is responsible for formatting a message to be injected into the network and sent back to the original requester. A network interface employing MSHRs and TSHRs is similar to the design utilized by the SGI Origin [215].

2.1.7   BRIEF STATE-OF-THE-ART SURVEY

Interaction of on-chip caches and on-chip network and memory system. Intel's Larrabee architecture [312] features a shared L2 cache design; each core has fast access to its own L2 subset and utilizes a ring network to communicate with remote L2 caches. The dynamic non-uniform cache architecture (NUCA) [182] utilizes an on-chip network among banks of a large shared cache to move data quickly to the processor. TRIPS employs both a scalar operand network and an on-chip cache network [138]. Traffic bound to and from memory represents a substantial fraction of overall NoC traffic and requires careful consideration [96, 317]. Memory controller placement can lead to network hotspots and must be carefully considered in large many-core systems [13]. NoC designs specifically tailored to memory traffic have also been proposed [112].

Interactions between cache coherence protocols and NoCs. To maintain a coherent view of memory, all nodes must observe memory requests in the same order. This ordering is typically achieved through implicit ordering intrinsic to the interconnect (e.g., bus, ring) or through the use of an ordering point (e.g., AMD Opteron [81]). Inherent ordering properties within the topology can also be beneficial for chip multiprocessor. Work by Marty and Hill [240] exploits the partial ordering of a ring to ease the implementation of coherence solutions. Support for multicast, broadcast and collective operations has become increasingly important considering their important role in coherence protocols [114, 206, 234]. Cache coherence protocols that leverage on-chip network properties or are embedded in-network have been explored as a way to further optimize communication energy and performance [15, 16, 111, 115], with the MIT SCORPIO chip (Chapter 8) demonstrating in-network ordering.

Traffic characterization and prioritization. Understanding NoC traffic characteristics is critical to facilitating traffic prioritization and designing quality of service (QoS) mechanisms. Recent work considers the relative importance of different coherence messages, bandwidth vs. latency sensitivity of messages and system-level behavior [98, 317, 375]. Quality of service mechanisms for on-chip networks have been studied [142, 217]. Tailored handling of different message sizes leads to improved performance and energy-efficiency in cache coherent on-chip networks which are typically characterized by having bimodal distribution of short control messages and long data messages [68, 226, 235, 237]. Performance isolation of traffic from different applications in the context of server consolidation workloads has also been explored [233, 236].

2.2   MESSAGE PASSING

The message passing paradigm requires explicit communication between processes. User communication is performed through operating system and library calls. Software must be written with matching send and receive calls to facilitate data transfer from one process to another. Through message passing, communication and synchronization between an arbitrary set of cooperating processes can be achieved.

Here we focus on the relationship between message passing and network design. In a shared memory paradigm, identifying or naming shared data is easily enabled by one globally shared address space. In message passing, the owning process of the data must be identified in order to request the data. Message types and sizes are very flexible in message passing. This can lead to a lot of overhead; it falls on software to decode and process the messages. Flexible message lengths can also complicate buffer management. Interrupts may be required so that software can temporarily store messages. In shared memory, storage of messages throughout the network and at the receiving processor is transparent to software and is completely hardware managed. Message passing attempts to amortize overheads and latencies associated with communication by communicating large chunks of data. The hardware cost and design complexity of message passing are generally considered lower than that of shared memory; implementing and verifying cache coherence protocols bring tremendous complexity to the design process. However, there are numerous trade-offs and message passing does introduce additional complexities elsewhere in the system.

Communication performance in message passing is often easier to model and reason about since communication happens explicitly. Programmers have clear guidelines and understanding of the cost of their communication; namely, messages are expensive so they should be sent infrequently. Shared memory is more challenging as communication occurs implicitly both through loads and stores but also through cache conflicts that will require additional communication not obvious at the software level.

Blocking vs. non-blocking. Blocking or synchronous message passing requires the sender to stall until the receiver has acknowledged the message. Although conceptually simple, blocking message passing must carefully account for deadlock, e.g., two processes issue send commands and stall. Neither process is able to proceed to the receive command and will wait indefinitely. Non-blocking or asynchronous message passing allows the sender to proceed immediately after sending the message. This removes deadlock-related complications but leads to additional complexities in storing messages until the receiver is ready to process them.

Message storage. Several different strategies can be employed for sending and storing messages. Messages can be written directly to dedicated registers or message buffers or can be stored in memory via memory mapped I/O. Receiving processors can be notified on messages via interrupts or by polling on memory-mapped locations. We will explore the various strategies employed by message-passing on-chip network chips in Chapter 8.

2.2.1 BRIEF STATE-OF-THE-ART SURVEY

In addition to the emergence of on-chip networks for cache-coherent chip multiprocessors, tiled microprocessors also leverage on-chip networks for scalar operand networks. A tiled microprocessor distributes functional units across multiple tiles; these designs mitigate problems of wire delay present in large superscalar architectures. Instructions are scheduled across available tiles. Architectures such as TRIPS [308], RAW [335], and Wavescalar [331] use operand networks to communicate register values between producing and consuming instructions. The result of an instruction is communicated to consumer tiles which can then wake up and fire instructions waiting on the new data. While we focus primarily on shared-memory CMPs here, several interconnection networks rely on message passing, including on-chip [145, 158, 335, 356] and off-chip designs [91, 127, 207, 221]. Additional details are given in Chapter 8.

2.3 NOC INTERFACE STANDARDS

On-chip networks have to adhere to standardized protocols so that they can plug-and-play with IP blocks that were also designed to interface with the same standard. Such standardized protocols define the rules for all signaling between the IP blocks and the communication fabric, while permitting configuration of specific instances. Several widely used standards for on-chip communications in SoCs today are ARM's AMBA [27], ST Microelectronics' STBus [246], Sonics' OCP [324], and OpenCores Wishbone [358]. Here, we will discuss some features that are common across these standards, using ARM's AMBA AXI [26] as the specific protocol for illustration. We refer interested readers to Pasricha and Dutt [289] that goes through all existing on-chip bus-based protocols in detail.

Bus-based transaction semantics. First, as current SoCs predominantly use buses as the on-chip interconnects, these standard interfaces have bus-based semantics where nodes connected to the interconnect are defined as masters or slaves, and communicate via transactions. Masters start a transaction by issuing requests; slaves then receive and subsequently process the request. The transaction is completed when the slave responds to the original request. This request-response transaction model matches those used in buses, making it easier to design network interface wrappers around IP blocks that were originally designed to interface with buses. For instance, a processor core will be a master that initiates a new transaction through issuing a write request to a memory module, while the memory module will be the slave that executes the write request and responds with an acknowledgment response.

Every transaction in the AMBA AXI protocol sends address and control information on the address channel, while data are sent on the data channel in bursts. Writes have an additional response channel. These channels are illustrated for AXI reads and writes in Figure 2.10. The sizes of these channels can range from 8 to 1024 bits, with a particular size instantiated for each design. So an NoC that interfaces using the AXI protocol has to have these three channels. A write from the master node will lead to its network interface encapsulating and translating the address in the address channel to the slave's node destination address in the message header, and the data in the write data channel encoded as the body of the message. The message will then be broken down into packets and injected into the injection port of the attached router. At the destination, the packets will be assembled into a message, and the address and control information extracted from the header and fed into the AXI write address channel, while the data are obtained from the body and fed into the AXI write data channel. Upon receipt of the last flit of the message, the network interface will then compose a write response message and send it back to the master node, feeding into the AXI write response channel at the master.

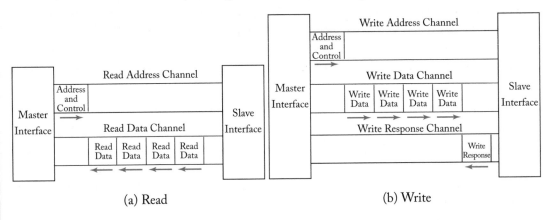

(a) Read   (b) Write

Figure 2.10: AXI read and write channels [26].

Out-of-order transactions. Many of the latest versions of these standards relax the strict ordering of bus-based semantics so point-to-point interconnect fabrics such as crossbars and on-chip networks can be plugged in, while retaining backward compatibility to buses, such as OCP 3.0 [297], AXI [26], and STNoC [245].

For instance, AXI relaxes the ordering between requests and responses, so responses need not return in the same order as that of requests. Figure 2.11 illustrates this feature of AXI which allows multiple requests to be outstanding and slaves to be operating at different speeds. This allows multiple address and data buses to be used, as well as split-transaction buses (where a

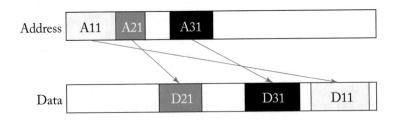

Figure 2.11: The AXI protocol allows messages to complete out of order: D21 returns data prior to D11 even though A11 occurred prior to A21.

transaction does not hold on to the bus throughout, but instead, requests and responses of the same transaction separately arbitrate for the bus), and ultimately, on-chip networks. In on-chip networks, packets sent between different pairs of nodes can arrive in different order from the sending order, depending on the distance between the nodes and the actual congestion level. A global ordering between all nodes is difficult to enforce. So out-of-order communication standards are necessary for on-chip network deployment.

Coherence. System-wide coherence support is provided by AMBA 4 ACE (AXI Coherency Extensions) and the more recent AMBA 5 CHI (Coherent Hub Interface) [26]. This is in the form of additional channels to support various coherence messages, snoop response controllers, barrier support, and QoS. This allows multiple processors to share memory for architectures like ARM's big.LITTLE.

2.4   CONCLUSION

This chapter introduces several system-level concepts that provide an important foundation and context for our discussion of on-chip networks. We provide a high level overview of how various architectural choices can impact on-chip network traffic. We also present a brief overview of interface standards. We will revisit the impact of architectural design choices in Chapter 7 on evaluation and in Chapter 8 where we present case studies of recent academic and industrial designs that feature on-chip networks.

CHAPTER 3

Topology

The on-chip network topology determines the physical layout and connections between nodes and channels in the network. The effect of a topology on overall network cost-performance is profound. A topology determines the number of hops (or routers) a message must traverse as well as the interconnect lengths between hops, thus influencing network latency significantly. As traversing routers and links incurs energy, a topology's effect on hop count also directly affects network energy consumption. Furthermore, the topology dictates the total number of alternate paths between nodes, affecting how well the network can spread out traffic and hence support bandwidth requirements. The implementation complexity cost of a topology depends on two factors: the number of links at each node (node degree) and the ease of laying out a topology on a chip (wire lengths and the number of metal layers required).

One of the simplest topologies is a bus which connects a set of components with a single, shared channel. Each message on the bus can be observed by all components on the bus; it is an effective broadcast medium. However, buses have limited scalability due to saturation of the shared channel as additional components are added.

In this chapter, we will focus on switched topologies, where a set of components is connected to one another via a set of routers and links. We first describe several metrics that are very useful for developing back-of-the-envelope intuition when comparing topologies. Next, we will describe several commonly used topologies in on-chip networks and compare them using these metrics.

3.1 METRICS

Since the first decision designers have to make when building an on-chip network is, frequently, the choice of the topology, it is useful to have a means for quick comparisons of different topologies before the other aspects of a network (such as its routing, flow control and microarchitecture) are even determined. Here, we describe several abstract metrics that come in handy when comparing different topologies. Figure 3.1 shows three commonly used on-chip topologies used to illustrate these metrics.

---

[1]Note that the figure illustrates the 2-D version of meshes and tori.

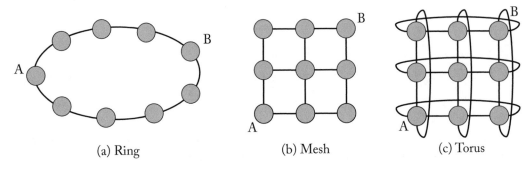

(a) Ring   (b) Mesh   (c) Torus

Figure 3.1: Common on-chip network topologies.[1]

3.1.1   TRAFFIC-INDEPENDENT METRICS

We first define a set of design-time metrics that are agnostic to the traffic flowing through the network.

degree

Degree. The degree of a topology refers to the number of links at each node. For instance, for the topologies in Figure 3.1, a ring topology has a degree of two since there are two links at each node, while a torus has a degree of four as each node has four links connecting it to four neighboring nodes. Note that in the mesh network, not all switches have a uniform degree. Degree is useful as a proxy for the network's cost, as a higher degree requires more ports at routers, which increases implementation complexity and adds area/energy overhead at each router. The number of ports per router is referred to as the router radix.

bisection bandwidth

Bisection bandwidth. The bisection bandwidth is the bandwidth across a cut that partitions the network into two equal parts.[2] For example, in Figure 3.1, two links cross the bisection for the ring, three for the mesh and six for the torus. This bandwidth is often useful in defining worst-case performance of a particular network, since it limits the total data that can be moved from one side of the system to the other. It also serves as a proxy for cost since it represents the amount of global wiring that will be necessary to implement the network. As a metric, bisection bandwidth is less useful for on-chip networks as opposed to off-chip networks, since global on-chip wiring is considered abundant relative to off-chip pin bandwidth.

diameter

Diameter. The diameter of the network is the maximum distance between any two nodes in the topology, where distance is the number of links in the shortest route. For example, in Figure 3.1 the ring has a diameter of four, the mesh has a diameter of four and the torus has a diameter of two. The diameter serves as a proxy for the maximum latency in the topology, in the absence of contention.

[2]If there are multiple such cuts possible, it is the minimum among all the cuts.

3.1.2   TRAFFIC-DEPENDENT METRICS

Next, we define a set of metrics that depends on the traffic (i.e., source-destination pairs) flowing through the network.

Hop count. The number of hops a message takes from source to destination, or the number of links it traverses, defines hop count. This is a very simple and useful proxy for network latency, since every node and link incurs some propagation delay, even when there is no contention. The maximum hop count is given by the *diameter* of the network. In addition to the maximum hop count, average hop count is very useful as a proxy for network latency. It is given by the average hops over all possible source-destination pairs in the network.

For the same number of nodes, and assuming uniform random traffic where every node has an equal probability of sending to every other node, a ring (Figure 3.1a) will lead to higher hop count than a mesh (Figure 3.1b) or a torus [93] (Figure 3.1c). For instance, in the figure shown, assuming bidirectional links and shortest-path routing, the maximum hop count of the ring is four, that of a mesh is also four, while a torus improves the hop count to two. Looking at average hop count, we see that the torus again has the lowest average hop count ($1\frac{1}{3}$). The mesh has a higher average hop count of $1\frac{7}{9}$. Finally, the ring has the worst average hop count of the three topologies in Figure 3.1 with an average of $2\frac{2}{9}$. The formulas for deriving these values will be presented in Section 3.2.

Maximum channel load. This metric is useful as a proxy for estimating the maximum bandwidth the network can support, or the maximum number of bits per second (bps) that can be injected by every node into the network before it saturates:

Maximum Injection Bandwidth = 1 / Maximum Channel Load.

Intuitively, it involves first determining which link or channel[3] in the network will be the most congested given a particular traffic pattern, as this link will limit the overall network bandwidth. For uniform random traffic, this link is often on the bisection cut. Next, the load on this channel is estimated. Since at this early stage of design, we do not yet know the specifics of the links we are using (how many actual interconnects form each channel, and each interconnects' bandwidth in bps), we need a relative way of measuring load. Here, we define it as being relative to the injection bandwidth. So, when we say the load on a channel is two, it means that the channel is loaded with twice the injection bandwidth. So, if we inject a flit every cycle at every node into the network, two flits will wish to traverse this specific channel every cycle. If the bottleneck channel can handle only one flit per cycle, it constrains the maximum bandwidth of the network to half the link bandwidth, i.e., at most, a flit can be injected every other cycle. Thus, the higher the maximum channel load, the lower the network bandwidth.

---

[3]We use link to refer to the physical set of wires connecting routers in an on-chip network, and channel to refer to the logical connection between routers in the network. In most designs, the link and channel are identical and can be used interchangeably.

Channel load can be calculated in a variety of ways, typically using probabilistic analysis. If routing and flow control are not yet determined, channel load can still be calculated assuming ideal routing (the routing protocol distributes traffic amongst all possible shortest paths evenly) and ideal flow control (the flow control protocol uses every cycle of the link whenever there is traffic destined for that link).

Here, we will illustrate this with a simple example, but the rest of the chapter will just show formulas for the maximum channel load of various common on-chip network topologies rather than walking through their derivations. Figure 3.2 shows an example network topology with two rings connected with a single channel. First, we assume uniform random traffic where every node has an equal probability of sending to every other node in the network including itself. To calculate maximum channel load, we need to first identify the bottleneck channel. Here, it is the single channel between the rings, shown in bold. We will assume it is a bidirectional link. With ideal routing, half of every node's injected traffic will remain within its ring, while the other half will be crossing the bottleneck channel. For instance, for every packet injected by node A, there is 1/8 probability of it going to either B, C, D, E, F, G, H, or itself. When the packet is destined for A, B, C, D, it does not traverse the bottleneck channel; when it is destined for E, F, G, H, it does. Therefore, 1/2 of the injection bandwidth of A crosses the channel. So does 1/2 of the injection bandwidth of the other nodes. Hence, the channel load on this bottleneck channel is 2. As a result the network saturates at 1/2 the injection bandwidth. Adding more nodes to both rings will further increase the channel load, and thus decrease the bandwidth.

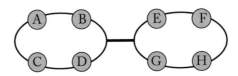

Figure 3.2: Channel load example with two rings connected via a single channel.

Path diversity. A topology that provides multiple shortest paths ($|R_{src-dst}| > 1$, where R represents the path diversity) between a given source and destination pair has greater path diversity than a topology where there is only a single path between a source and destination pair ($|R_{src-dst}| = 1$). Path diversity within the topology gives the routing algorithm more flexibility to load-balance traffic which reduces channel load and thus increases throughput. Path diversity also enables packets to potentially route around faults in the network. The ring in Figure 3.1a provides no path diversity ($|R| = 1$), because there is only one shortest path between pairs of nodes. If a packet travels clock-wise between A and B (in Figure 3.1a), it traverses four hops; if the packet goes counter-clockwise, it traverses five hops. More paths can be supplied only at the expense of a greater distance traveled. With an even number of nodes in a ring, two nodes that are half-way around the ring from each other will have a path diversity of two due to two minimal paths. On the other hand, the mesh and torus in Figures 3.1b and c provide a wider selection of distinct paths between source and destination pairs. In Figure 3.1b, the mesh supplies six distinct paths between A and B, all at the shortest distance of four hops.

3.2   DIRECT TOPOLOGIES: RINGS, MESHES, AND TORI

A direct network is one where each terminal node (e.g., a processor core or cache in a chip multiprocessor) is associated with a router; all routers act as both sources/sinks of traffic and as switches for traffic from other nodes. To date, most designs of on-chip networks have used direct networks since co-locating routers with terminal nodes is often most suitable in area-constrained environments on a chip.

direct network

Direct topologies can be described as k-ary n-cubes, where k is the number of nodes along each dimension, and n is the number of dimensions. For instance, a 4×4 mesh or torus is a 4-ary 2-cube with 16 nodes, a 8×8 mesh or torus is a 8-ary 2-cube with 64 nodes, while a $4 \times 4 \times 4$ mesh or torus is a 4-ary 3-cube with 64 nodes. This notation assumes the same number of nodes on each dimension, so total number of nodes is k^n. Practically speaking, most on-chip networks utilize 2-D topologies that map well to the planar substrate as otherwise, more metal layers will be needed; this is not the case for off-chip networks where cables between chassis provide 3-D connectivity. In each dimension, k nodes are connected with channels to their nearest neighbors. Rings fall into the torus family of network topologies as k-ary 1-cubes.

k-ary n-cubes

With a torus, all nodes have the same degree; however, with a mesh, nodes along the edge of the network have a lower degree than nodes in the center of the network. A torus is also edge-symmetric (a mesh is not), this property helps the torus network balance traffic across channels. Due to the absence of edge-symmetry, a mesh network experiences significantly higher demand for center channels than for edge channels.

edge symmetry

Next, we examine the values for the torus and the mesh in terms of the abstract metrics given in Section 3.1. A torus network requires two channels in each dimension or $2n$. So for a 2-D torus, the degree would be four and for a 3-D torus, the degree would be six. The degree is the same for a mesh, although some ports on the edge of the network will go unused. The average hop count for a torus network is found by averaging the minimum distance between all possible node pairs. This gives

$$H_{avg} = \begin{cases} \frac{nk}{4} & k\ even \\ n(\frac{k}{4} - \frac{1}{4k}) & k\ odd \end{cases}.$$

Without the wrap-around links of a torus, the average minimum hop count for a mesh is slightly higher and is given by

$$H_{avg} = \begin{cases} \frac{nk}{3} & k\ even \\ n(\frac{k}{3} - \frac{1}{3k}) & k\ odd \end{cases}.$$

The maximum channel load across the bisection of a torus under uniform random traffic with an even k is $k/8$ limiting the maximum injection throughput to $8/k$ flits/node/cycle. For a mesh, the channel load increases to $k/4$ and lowers the maximum injection throughput to $4/k$ flits/node/cycle.

Both mesh and torus networks provide path diversity for routing messages, compared to a ring. As the number of dimensions increase, so does the path diversity.

3.3   INDIRECT TOPOLOGIES: CROSSBARS, BUTTERFLIES, CLOS NETWORKS, AND FAT TREES

Indirect networks connect terminal nodes via one or more intermediate stages of switch nodes. Only terminal nodes are sources and destinations of traffic, intermediate nodes simply switch traffic to and from terminal nodes.

The simplest indirect topology is known as a crossbar. A crossbar connects n inputs to m outputs via $n \times m$ simple crosspoint switch nodes. It is called *non-blocking* as it can always connect a sender to a unique receiver. Crossbars will be discussed further as components of router microarchitectures in Chapter 6.

The butterfly network is an example of an indirect topology. Butterfly networks can be described as k-ary n-flies. Such a network would consist of k^n terminal nodes (e.g., cores, mem-

ory), and comprises n stages of k^{n-1} $k \times k$ intermediate switch nodes. In other words, k is the degree of the switches, and n is the number of stages of switches. Figure 3.3 illustrates a 2-ary 3-fly network; source and destination nodes are shown as logically separate in this figure with source nodes on the left and destination nodes on the right.

Next, we analyze the butterfly using the metrics given in Section 3.1. The degree of each intermediate switch in a butterfly network is given as $2k$. Unlike the mesh or the torus where the hop count varies based on source-destination pair, every source-destination pair in a butterfly network experiences the same hop count given by $n-1$ (assuming that the source and destination nodes are also switches). With uniformly distributed traffic, the maximum channel load for the butterfly is 1, leading to a peak injection throughput of 1 flit/node/cycle. Other traffic patterns that require significant traffic to be sent from one half of the network to the other half will increase the maximum channel load, lowering the injection throughput.

The primary disadvantages of a butterfly network are the lack of path diversity and the inability of these networks to exploit locality. With no path diversity, a butterfly network performs poorly in the face of unbalanced traffic patterns such as when each node in one half of the network sends messages to a node in the other half.

A folded version of the butterfly, known as Flattened Butterfly [186] folds all intermediate switches along a row into one switch, converting the indirect version of the topology to a direct version. Each 2×2 switch now becomes a higher-radix switch. The 4×4 version is shown in Figure 3.4 where every router has 7 ports (including the one from the core which is not shown in the figure). Each destination can be reached with a maximum of two hops. However, minimal routing can do a poor job of balancing the traffic load, so non-minimal paths have to be selected, thereby increasing the hop count.

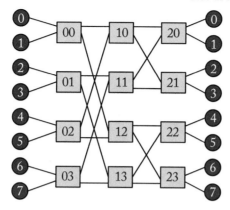

Figure 3.3: A 2-ary 3-fly butterfly network.

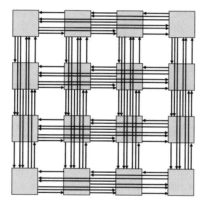

Figure 3.4: A 4 × 4 flattened butterfly network.

A symmetric Clos network is a three-stage[4] network characterized by the triple, (m, n, r) where m is the number of middle stage switches, n is the number of input/output ports on each input/output switch (first and last stage switches), and r is the number of first/last stage switches. When $m > 2n - 1$, a Clos network is strictly non-blocking, i.e., any input port can connect to any unique output port, like a crossbar. A Clos network consists of $r \times n$ nodes. A 3-stage Clos has a hop count of four for all source destination pairs. A Clos network does not use identical switches at each stage. The degree of the first and last stage switches is given by $n + m$ while the degree of the middle switches is $2r$. With m middle stages, a Clos network provides path diversity of $|R_{src-dst}| = m$. A disadvantage of a Clos network is its inability to exploit locality

clos

----

[4]A Clos network with a larger number of odd stages can be built by recursively replacing the middle switches with a 3-stage Clos.

between source and destination pairs. Figure 3.5 depicts a 3-stage Clos network characterized by the triple $(5, 3, 4)$.

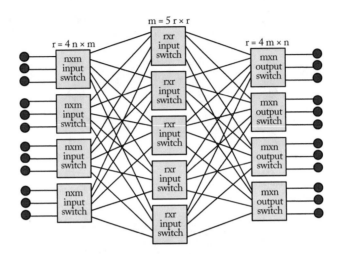

Figure 3.5: An $(m = 5, n = 3, r = 4)$ symmetric Clos network with $r = 4n \times m$ input-stage switches, $m = 5r \times r$ middle-stage switches, and $r = 4m \times n$ output-stage switches. Crossbars form all switches.

A Clos network can be folded along the middle set of switches so that the input and output switches are shared. In Figure 3.6b, a 5-stage folded Clos network characterized by the triple $(2, 2, 4)$ is depicted. The center stage is realized with another 3-stage Clos formed using $(2, 2, 2)$ Clos network. This Clos network is folded along the top row of switches.

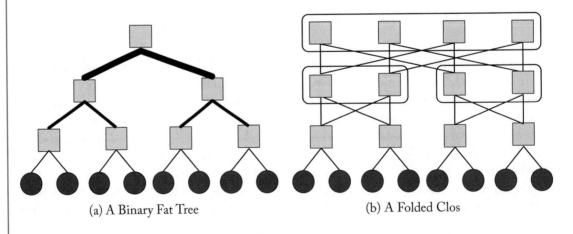

(a) A Binary Fat Tree   (b) A Folded Clos

Figure 3.6: A fat tree network.

A fat tree [222] is logically a binary tree network in which wiring resources increase for stages closer to the root node (Figure 3.6a). A fat tree can be constructed from a folded Clos network, as shown in Figure 3.6b giving path diversity over the tree network in Figure 3.6a. The Clos is folded back on itself at the root, logically giving a 5-stage Clos network. In a fat tree, messages are routed up the tree until a common ancestor is reached and then routed down to the destination; this allows the fat tree to take advantage of locality between communicating nodes. Each switch in the fat tree has a logical degree of four, although the links in higher-level nodes are much wider than those in the lower levels.

fat tree

3.4   IRREGULAR TOPOLOGIES

MPSoC design may leverage a wide variety of heterogeneous IP blocks; as a result of the heterogeneity, regular topologies such as a mesh or a torus described above may not be appropriate. With these heterogeneous cores, a customized topology will often be more power efficient and deliver better performance than a standard topology.

Often, communication requirements of MPSoCs are known *a priori*. Based on these structured communication patterns, an application characterization graph can be constructed to capture the point-to-point communication requirements of the IP blocks. To begin constructing the required topology, the number of components, their size, and their required connectivity as dictated by the communication patterns must be determined.

An example of a customized topology for a video object plane decoder is shown in Figure 3.7. The MPSoC is composed of 12 heterogeneous IP blocks. In Figure 3.7a, the design is mapped to a 3×4 mesh topology requiring 12 routers (R). When specific application characteristics are taken into account (e.g., not every block needs to communicate directly with every other block), a custom topology is created (Figure 3.7b). This irregular topology reduces the number of switches from 12 to 5; by reducing the number of switches and the links in the topology, significant power and area savings are achieved. Some blocks can be directly connected without the need for a switch, such as the VLD and run length decoder units. Finally, the degree of the switches has changed; the mesh in Figure 3.7a requires a switch with 5 input/output ports (although ports can be trimmed on edge nodes). The 5 input/output ports represent the four cardinal directions: north, south, east and west plus an Injection/Ejection port. All of these ports require both input and output connections leading to 5×5 crossbars. With a customized topology, not all blocks need both input and output ports; the largest switch in Figure 3.7b is a 4×4 switch. Not every connection between links coming into and out of a router is necessary in the customized topology resulting in smaller switches; connectivity has been limited because full connectivity is not needed by this specific application.

3.4.1   SPLITTING AND MERGING

Two types of techniques have been explored for customizing a topology: splitting and merging. With splitting, a large crossbar connecting all nodes is first created and then iteratively split into

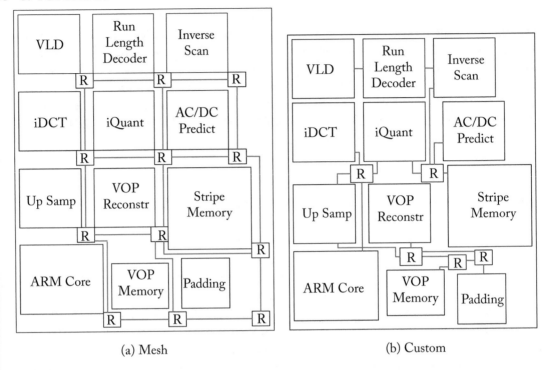

Figure 3.7: A regular (mesh) topology and a custom topology for a video object plane decoder (VOPD) (from [47]).

multiple small switches to accommodate a set of design constraints. Alternatively, a network with a larger number of switches such as a mesh or torus can be used as a starting point. From this starting point, switches are merged together to reduce area and power.

Splitting. One technique for arriving at a customized network topology is to begin with a large fully connected switch (crossbar). Such a large crossbar will likely violate the design constraints and must iteratively be split into smaller switches until design constraints are satisfied [157]. When a switch is split into two smaller switches creating a partition, the bandwidth provided between the two switches must satisfy the volume of communication that must now flow between partitions. Nodes can be moved between partitions to optimize the volume of communication between switches.

Merging. An alternative to iteratively splitting larger switches into smaller switches is to begin with large number of switches and merge them [295, 326]. By merging adjacent routers in the topology, power and area costs can be reduced. In this type of design flow, floorplanning of the various MPSoC components is done as the first step. Floorplanning can be done based on the application characterization graph, e.g., nodes that communicate heavily should be placed in close proximity during floor planning. Next, routers are placed at each of the channel intersection points, where three or more channels merge or diverge. The last step merges adjacent routers if they are close together and if merging will not violate bandwidth or performance constraints and given that there is some benefit from such merging (e.g., power is reduced).

3.4.2 TOPOLOGY SYNTHESIS ALGORITHM EXAMPLE

Topology synthesis and mapping is an NP-hard problem [295]; as a result, several heuristics have been proposed to find the best topology in an efficient manner. In this section, we provide an example of one such application-specific topology synthesis algorithm for a MPSoC from Murali et al. [258]. This algorithm is an example of a splitting algorithm; they begin with an application communication graph showing the bandwidth required between the various application tasks as shown in Figure 3.8a.

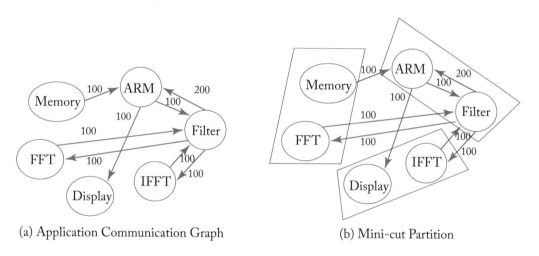

(a) Application Communication Graph   (b) Mini-cut Partition

Figure 3.8: Topology synthesis algorithm example.

The algorithm synthesizes a number of different topologies, starting with a topology where all IP cores are connected through one large switch to the other extreme where each core has its own switch. For each switch count, the algorithm tunes the operating frequency and the link width. For a given switch count i, the input graph (Figure 3.8a) is partitioned into i min-cut partitions. Figure 3.8b shows a min-cut partition for $i = 3$. The min-cut partition is performed so that the edges of the graph that cross partitions have lower weights than the edges within partitions. Additionally, the number of nodes assigned to each partition remains nearly the same. Such a min-cut partition will ensure that traffic flows that have high bandwidth use the same switch for communication.

Once the min-cut partitions have been determined, routes must be restricted to avoid deadlocks. We discuss deadlock avoidance in Chapter 4. Next, physical links between switches

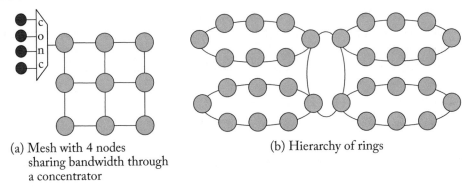

(a) Mesh with 4 nodes sharing bandwidth through a concentrator

(b) Hierarchy of rings

Figure 3.9: Hierarchical topologies.

must be established and paths must be found for all traffic flows through the switches. Once the size of the switches and their connectivity is determined, the design can be evaluated to see if power consumption of the switches and hop count objectives have been met. Finally, a floorplanner is used to determine the area and wire lengths of a synthesized design.

3.5   HIERARCHICAL TOPOLOGIES

Up to this point, we have assumed a one-to-one correspondence between network nodes and terminal nodes. We have also assumed a uniform topology across the entire system. However, these need not be the case. In real systems, multiple nodes might be clustered together in one topology, and these clusters connected together via another topology, building a hierarchical design.

The simplest form of a hierarchical topology is one where multiple cores share the same router node using concentrators. Figure 3.9a shows such a concentrated mesh, where four terminal nodes (cores, caches, etc) share a network router. The use of concentration reduces the number of routers needed in the network, thereby reducing the hop count and the size (area) of the network. This also helps scale networks to larger sizes. In Figure 3.9a concentration allows a 3×3 mesh to connect 36 nodes with only 9 routers, instead of 36. However, on the flip side, concentration can increase network complexity. The concentrator must implement a policy for sharing injection bandwidth. This policy can dynamically share bandwidth or statically partition bandwidth so that each node gets $\frac{1}{c}$ the injection bandwidth, where c is the concentration factor. Another drawback of using concentration is that during periods of bursty communication, the injection port bandwidth can become a bottleneck.

Another hierarchical topology is shown in Figure 3.9b. A 32-core chip is partitioned into 8 clusters. Each cluster is built by connecting eight cores with a bi-directional ring. The four

concentration

rings are connected together via another ring. The challenge in such a hierarchical topology is the arbitration for bandwidth into the central ring.

3.6   IMPLEMENTATION

In this section, we discuss the implementation of topologies on a chip, looking at both physical layout implications, and the role of abstract metrics defined at the beginning of the chapter.

3.6.1   PLACE-AND-ROUTE

There are two components of the topology which require careful thought during the physical design: links and routers.

The links are routed on semi-global or global metal layers, depending on the channel widths and the distance they need to traverse. Wire capacitance tends to be an order of magnitude higher than that of transistors, and can dominate the energy of the network if not optimized well. The target clock frequency determines the size of and distance between repeaters[5] that need to be inserted to meet timing. Thicker wires with larger spacing between wires can be employed to lower the wire resistance and coupling capacitance, thus increasing speed and energy efficiency. However, metal layer density rules and design rules checking (DRC) can limit how much one can play around with these parameters. In terms of area, the core to core links can be routed over active logic, mitigating any area overheads apart from those of the repeaters. But care needs to be taken since active switching of transistors can introduce cross talk in the wires. Similarly, routing toggling links over sensitive circuits such as SRAMs which operate at low voltages could introduce glitches and errors, leading to the area over caches usually being blocked from wiring. Hence, the floorplanning of the entire chip needs to carefully consider where router links lie relative to processor cores, caches, memory controllers, etc.

When implementing routers, the node degree (i.e., the number of ports in and out of the router) determines the overhead, since each port has associated buffering and state logic, and requires a link to the next node. As a result, while rings have poorer network performance (latency, throughput, energy and reliability) when compared to higher-dimensional networks, they have lower implementation overhead as they have a node degree of two while a mesh or torus has a node degree of four. Similarly, high-radix topologies such as the 4×4 flattened butterfly discussed in Section 3.3 have lower latency and higher throughput than a mesh for the same channel width, but the seven ported routers add a higher area and energy footprint, especially due to the larger crossbar switch whose area grows as a square of the number of ports.

The 2-D floorplan of the logical topology can also often lead to implementation overheads. As an example, the torus from Figure 3.1 has to be physically arranged in a folded form to equalize wire lengths (see Figure 3.10) instead of employing long wrap-around links between edge nodes. As a result, wire lengths in a folded torus are twice that in a mesh of the same size, so

folded torus

---

[5]An inverter or a pair of inverters.

per-hop latency and energy are actually higher. Furthermore, a torus requires twice the number of links which must be factored into the wiring budget. If the available wire tracks along the bisection is fixed, a torus will be restricted to narrower links than a mesh, thus lowering per-link bandwidth, and increasing transmission delay. From an architectural comparison on the other hand, a torus has lower hop count (which leads to lower delay and energy) compared to a mesh. These contrasting properties illustrate the importance of considering implementation details in selecting between alternative topologies.

Similarly, trying to create an irregular topology optimized for an application's communication graph could end up having many criss-crossing links. These would show up as wire congestion during place-and-route forcing the automated tools or the designer to route around congested nets adding delay and energy overheads.

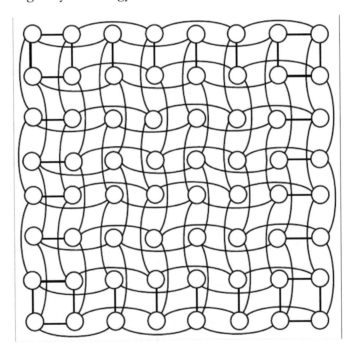

Figure 3.10: Layout of a 8 × 8 folded torus.

3.6.2   IMPLICATION OF ABSTRACT METRICS

We introduced various abstract metrics at the beginning of this chapter, and used them to compare and contrast various common topologies. Here, we will discuss the implications of these simple metrics on on-chip network implementation, explaining why they are good proxies for on-chip network delay, area, and power, while highlighting common pitfalls in the use of these metrics.

Node degree is useful as a proxy for router complexity, as higher degree implies greater port count. Adding a port in a router leads to additional input buffer queue(s), additional requestors to the allocators, as well as additional ports to the crossbar switch, all major contributors to a router's critical path delay, area footprint, and power. While router complexity is definitely increased for topologies with higher node degree, link complexity does not correlate directly with node degree. This is because link complexity depends on the link width, as link area and power overheads correlate more closely with the number of wires than the number of ports. So if the same number of wires is divided amongst a 2-port router and a 3-port router, link complexity will be roughly equal.

Hop count is a metric that is widely used as a proxy for overall network latency and power. Intuitively, it makes sense, since flits typically have to stop at each hop, going through the router pipeline followed by the link delay. However, hop count does not always correlate with network latency in practice, as it depends heavily on the router pipeline length and the link propagation delay. For instance, a network with only two hops, a router pipeline depth of 5 cycles, and long inter-router distances requiring 4 cycles for link traversal, will have an actual network latency of 18 cycles. Conversely, a network with three hops where each router has a single-cycle pipeline and the link delay is a single cycle, will have a total network latency of only six cycles. If both networks have the same clock frequency, the latter network with the higher hop count will instead be faster. Unfortunately, factors such as router pipeline depth are typically not known until later in the design cycle.

With topologies typically trading off node degree and hop count, i.e., a topology may have low node degree but high average hop count (e.g., a ring), while another may have high node degree but low average hop count (e.g., a mesh), comparisons between topologies become trickier. Implementation details have to be factored in before an astute choice can be made.

Maximum channel load is another metric that is useful as a proxy of network performance and throughput. Here, it is a good proxy for network saturation throughput and maximum power. The higher the maximum channel load on a topology, the greater the congestion in the network caused by the topology and routing protocol, and thus, the lower the overall realizable throughput. Clearly, the specific traffic pattern affects maximum channel load substantially and representative traffic patterns should be used in estimating maximum channel load and through-put. Since it is a good proxy for saturation, it is also very useful for estimating peak power, as dynamic power is highest with peak switching activity and utilization in the network.

Bisection bandwidth is typically used as the metric to define the bandwidth of the net-work. The channel load on the bisection links determines the peak achievable throughput of the network for uniform random traffic. For instance, for a 8×8 mesh, the channel load on the bisection links, as discussed in Section 3.2, is $8/4 = 2$, setting the peak injection throughput to $1/2 = 0.5$ flits/node/cycle. However, the actual throughput achieved by the network will be lower than this due to imperfect load balance. Imperfect load balance results from inefficiencies in the routing and flow-control protocols which determine the actual amount of data that can use the bisection links every cycle; these issues will be discussed in later chapters.

3.7   BRIEF STATE-OF-THE-ART SURVEY

Significant research exists into all of the topologies discussed in this chapter. Much of this research has been conducted with respect to off-chip networks in the past [86, 110, 294], but fundamental concepts apply to on-chip networks as well.

The majority of on-chip network proposals gravitate toward either ring or mesh topologies. For example, the IBM Cell processor, the first product with an on-chip network, used a ring topology, largely for its design simplicity, its ordering properties and low power consumption. Four rings are used to boost the bandwidth and to help alleviate the latency (halving the average hop count). Similarly, the proposed Intel Larrabee [312] also adopted the two-ring topology. Another simple, regular topology, the mesh, has also been adopted. The MIT Raw chip, the first chip with an on-chip network has four meshes. Chapter 8 elaborates further.

In Balfour and Dally [38], a comparison of various on-chip network topologies including a mesh, concentrated mesh, torus, and fat tree, is presented. Furthermore, cost (including power and area) and performance are considered in this design space exploration. Balfour and Dally also suggest potential benefits for employing multiple parallel networks (all of the same topologies) to improve network throughput. Multiple meshes have been used in the MIT Raw chip and its follow-on commercialization into the Tilera TILE64 chip [356]. Each parallel network is a mesh topology and different types of traffic are routed over the distinct networks. Multiple mesh networks have also been proposed to improve energy efficiency of the network [100]. Multiple rings have been used in the IBM Cell [293] and the proposed Intel Larrabee [312].

Novel topologies have also been proposed for on-chip networks, focusing on the unique properties of on-chip networks such as the availability of large number of wiring tracks as well as the irregularity of MPSoC's traffic demands. Examples include the flattened butterfly [186], the dragonfly [187], a hierarchical star [218], the dodec [368], and the spidergon [83]. Hierarchical combinations of topologies, as well as the addition of express links between non-adjacent nodes have also been proposed [23, 88, 97, 141], and tailored to the MPSoC domain where there is prior knowledge of on-chip network bandwidth demands and connectivity [156, 268]. General purpose topologies tailored to specific characteristics of memory traffic have been proposed for GPUs [35] and for scale-out datacenter workloads [230]. Recent work also explores scaling on-chip network topologies to hundreds, thousands of cores [10, 142].

In the MPSoC domain, a variety of topologies have been explored. SPIN [22] proposes using a fat tree network. Both Æthereal [135] and xpipes [169] leverage irregular topologies customized for specific application demands. The Nostrum design relies on a mesh [247]. Bolotin et al. [53] propose trimming unnecessary links from a mesh and leveraging non-uniform link bandwidth within the mesh.

CHAPTER 4

Routing

After determining the network topology, the routing algorithm is used to decide what path a message will take through the network to reach its destination. The goal of the routing algorithm is to distribute traffic evenly among the paths supplied by the network topology, so as to avoid hotspots and minimize contention, thus improving network latency and throughput. All of these performance goals must be achieved while adhering to tight constraints on implementation complexity: routing circuitry can stretch critical path delay and add to a router's area footprint. While energy overhead of routing circuitry is typically low, the specific route chosen affects hop count directly, and thus substantially affects energy consumption. In addition, the path diversity enabled by the routing algorithm is also useful for increasing resiliency in the presence of network faults.

4.1 TYPES OF ROUTING ALGORITHMS

In this section, we briefly discuss various classes of routing algorithms. Routing algorithms are generally divided into three classes: deterministic, oblivious and adaptive.

While numerous routing algorithms have been proposed, the most commonly used routing algorithm in on-chip networks is dimension-ordered routing (DOR) due to its simplicity. Dimension-ordered routing is an example of a deterministic routing algorithm, in which all messages from node A to B will always traverse the same path. With DOR, a message traverses the network dimension-by-dimension, reaching the ordinate matching its destination before switching to the next dimension. In a 2-D topology such as the mesh in Figure 4.1, X-Y dimension-ordered routing sends packets along the X-dimension first, followed by the Y-dimension. A packet travelling from (0,0) to (2,3) will first traverse 2 hops along the X-dimension, arriving at (2,0), before traversing 3 hops along the Y-dimension to its destination. *dimension-order routing (DOR)* *deterministic routing*

Another class of routing algorithms are oblivious ones, where messages traverse different paths from A to B, but the path is selected without regard to network congestion. For instance, a router could randomly choose among alternative paths prior to sending a message. Figure 4.1 shows an example where messages from (0,0) to (2,3) can be randomly sent along either the Y-X route or the X-Y route. Deterministic routing is a subset of oblivious routing. *oblivious routing*

A more sophisticated routing algorithm can be adaptive, in which the path a message takes from A to B depends on network traffic situation. For instance, a message can be initially following the X-Y route and see congestion at (1,0)'s east outgoing link. Due to this congestion, *adaptive routing*

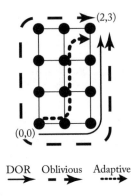

DOR Oblivious Adaptive

Figure 4.1: *DOR* illustrates an X-Y route from (0,0) to (2,3) in a mesh, while *Oblivious* shows two alternative routes (X-Y and Y-X) between the same source-destination pair that can be chosen obliviously prior to message transmission. *Adaptive* shows a possible adaptive route that branches away from the X-Y route if congestion is encountered at (1,0).

the message will instead choose to take the north outgoing link toward the destination (see Figure 4.1).

Routing algorithms can also be classified as minimal and non-minimal. Minimal routing algorithms select only paths that require the smallest number of hops between the source and the destination. Non-minimal routing algorithms allow paths to be selected that may increase the number of hops between the source and destination. In the absence of congestion, non-minimal routing increases latency and also power consumption as additional routers and links are traversed by a message. With congestion, the selection of a non-minimal route that avoids congested links, may result in lower latency for packets.

Before we get into details on specific deterministic, oblivious, and adaptive routing algorithms, we will discuss the potential for deadlock that can occur with a routing algorithm.

4.2 DEADLOCK AVOIDANCE

In selecting or designing a routing algorithm, not only must its effect on delay, energy, throughput and reliability be taken into account, most applications also require the network to guarantee deadlock freedom. A deadlock occurs when a knotted[1] cycle exists among the paths of multiple messages. Figure 4.2 shows four gridlocked (deadlocked) messages waiting for links that are currently held by other messages, preventing any message from making forward progress. The packet entering router A from the South input port is waiting to leave through the East output port, but another packet is holding onto that exact link while waiting at router B to leave via the

---

[1]In adaptive routing, cycles are necessary but not sufficient condition for deadlocks, as there could exist a cycle but there is a way out of this cycle, such as through an escape path. Knotted cycles more precisely define deadlock situations [347, 348].

minimal routing
non-minimal routing

routing deadlock

South output port, which is again held by another packet that is waiting at router C to leave via the West output port and so on.

Deadlock freedom can be ensured either in the routing algorithm, by preventing cycles among the routes generated by the algorithm, or in the the flow control protocol, by preventing router buffers from being acquired and held in a cyclic manner [110, 294]. The former will be discussed in this chapter, while the latter will be discussed in Chapter 5.6.

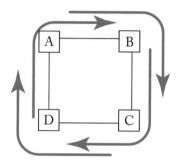

Figure 4.2: A classic network deadlock where four packets cannot make forward progress as they are waiting for links that other packets are holding on to.

4.3   DETERMINISTIC DIMENSION-ORDERED ROUTING

A routing algorithm can be described by which turns are permitted. Figure 4.3a illustrates all possible turns in a 2-D mesh network while Figure 4.3b illustrates the more limited set of permissible turns allowed by DOR X-Y routing. Allowing all turns results in cyclic resource dependencies, which can lead to network deadlock. To prevent these cyclic dependencies, certain turns should be disallowed. As you can see, no cycle is present in Figure 4.3b. Specifically, a message traveling east or west is allowed to turn north or south; however, messages traveling north and

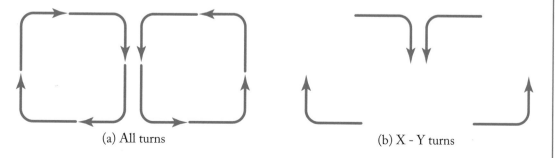

(a) All turns   (b) X - Y turns

Figure 4.3: Possible routing turns for a 2-D Mesh.

south are permitted no turns. Two of the four turns in Figure 4.2 will not be permitted, so a cycle is not possible.

Alternatively, Y-X routing can be used where messages traveling north or south are allowed to turn east or west but once a message is traveling East or West, no further turns are permitted. Depending on the network dimensions, i.e., whether there are more nodes along X or Y, one of these routing algorithms will balance load better with uniform random traffic since channel load is higher along the dimension with fewer nodes.

Dimension order routing is both simple and deadlock-free; however, it eliminates path diversity in a mesh network and thus lowers throughput. With dimension order routing, exactly one path exists between every source and destination pair. Without path diversity, the routing algorithm is unable to route around faults in the network or avoid areas of congestion. As a result of routing restrictions, dimension order routing does a poor job of load balancing the network.

4.4   OBLIVIOUS ROUTING

Using an oblivious routing algorithm, routing paths are chosen without regard to the state of the network. By not using information about the state of the network, these routing algorithms can be kept simple.

Valiant's routing algorithm

Valiant's randomized routing algorithm [342] is one example of an oblivious routing algorithm. To route a packet from source s to destination d using Valiant's algorithm, an intermediate destination d' is randomly selected. The packet is first routed from s to d' and then from d' to d. By routing first to a randomly selected intermediate destination before routing to the final destination, Valiant's algorithm is able to load balance traffic across the network; the randomization causes any traffic pattern to appear to be uniform random. Load balancing with Valiant's algorithm comes at the expense of locality; for example, by routing to an intermediate destination, the locality of near neighbor traffic on a mesh is destroyed. Hop count is increased, which in turn increases the average packet latency and the average energy consumed by the packet in the network. Besides, not only is locality destroyed with Valiant's algorithm, the maximum channel load can also be doubled, halving the network bandwidth.

Valiant's routing algorithm can be restricted to support only minimal routes [259], by restricting routing choices to only the shortest paths in order to preserve locality. In a k-ary n-cube topology, the intermediate node d' must lie within the minimal quadrant; the smallest n-dimensional sub-network with s and d as corner nodes bounding this quadrant.

With Valiant's routing whether considering minimal or non-minimal selection of d', dimension order routing can be used to route from s to d' and from d' to d. If DOR is used, not all paths will be exploited but better load balancing is achieved than deterministic routing from s directly to d. Figure 4.4 illustrates a routing path selected using Valiant's algorithm and minimal oblivious routing. In Figure 4.4a, Valiant's algorithm randomly selects an intermediate destination d'. The random selection can destroy locality and significantly increase hop count; here, the hop count is increased from three hops to nine hops. To preserve locality, minimal oblivious routing can be employed as in Figure 4.4b. Now, *d'* can only be selected to lie within the minimal quadrant formed by *s* and *d*, preserving the minimum hop count of three. One possible selection is highlighted (two other paths are possible for this source-destination pair as shown with dashed lines).

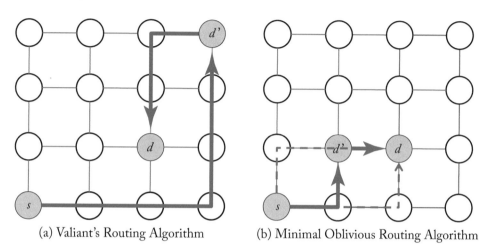

(a) Valiant's Routing Algorithm   (b) Minimal Oblivious Routing Algorithm

Figure 4.4: Oblivious routing examples.

Valiant's routing algorithm and minimal oblivious routing are deadlock free when used in conjunction with X-Y routing. An example of an oblivious routing algorithm that is not deadlock free is one that randomly chooses between X-Y or Y-X routes. The oblivious algorithm that randomly chooses between X-Y or Y-X routes is not deadlock-free because all four turns from Figure 4.2 are possible leading to potential cycles in the link acquisition graph.

4.5   ADAPTIVE ROUTING

A more sophisticated routing algorithm can be adaptive, i.e., the path a message takes from A to B depends on the network traffic situation. For instance, a message can be going along the X-Y route, see congestion at (1,0)'s east outgoing link and instead choose to take the north outgoing link toward the destination (see Figure 4.1).

Local or global information can be leveraged to make adaptive routing decisions. Adaptive routing algorithms often rely on local router information such as queue occupancy and queuing delay to gauge congestion and select links [89]. The backpressure mechanisms used by flow control (discussed in the next chapter) allow congestion information to propagate from the congestion site back through the network.

Figure 4.5 shows all possible (minimal) routes that a message can take from Node (0,0) to Node (2,3). There are nine possible paths. An adaptive routing algorithm that leverages only

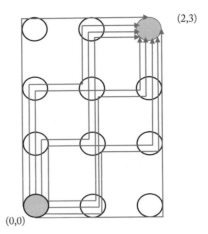

(2,3)

(0,0)

Figure 4.5: Adaptive routing example.

minimal paths could exploit a large degree of path diversity to provide load balancing and fault tolerance.

Adaptive routing can be restricted to taking minimal routes between the source and the destination. An alternative option is to employ misrouting, which allows a packet to be routed in a non-productive direction resulting in non-minimal paths. When misrouting is permitted, livelock becomes a concern. Without mechanisms to guarantee forward progress, livelock can occur as a packet is continuously misrouted so as to never reach its destination. We can combat this problem by allowing a maximum number of misroutes per packet and giving higher priority to packets than have been misrouted a large number of times. Misrouting increases the hop count but may reduce end-to-end packet latency by avoiding congestion (queueing delay).

With a fully adaptive routing algorithm, deadlock can become a problem. For example, the adaptive route shown in Figure 4.1 is a superset of oblivious routing and is subject to potential deadlock. Planar-adaptive routing [73] limits the resources needed to handle deadlock by restricting adaptivity to only two dimensions at a time. Duato has proposed flow control techniques that allow full routing adaptivity while ensuring freedom from deadlock [109]. Deadlock-free flow control will be discussed in Chapter 5.

Another challenge with adaptive routing is preserving inter-message ordering as may be needed by the coherence protocol. If messages must arrive at the destination in the same order that the source issued them, adaptive routing can be problematic. Mechanisms to re-order messages at the destination can be employed or messages of a given class can be restricted in their routing to prevent re-ordering.

misrouting

livelock

ADAPTIVE TURN MODEL ROUTING

We introduced turn model routing earlier in Section 4.3 and discussed how dimension order X-Y routing eliminates two out of four turns (Figure 4.3). Here, we explain how turn model can be more broadly applied to derive deadlock-free adaptive routing algorithms. Adaptive turn model routing eliminates the minimum set of turns needed to achieve deadlock freedom while retaining some path diversity and potential for adaptivity.

turn model routing

With dimension order routing only four possible turns are permitted of the eight turns available in a 2-D mesh. Turn model routing [131] increases the flexibility of the algorithm by allowing six out of eight turns. Only one turn from each cycle is eliminated.

In Figure 4.6, three possible routing algorithms are illustrated. Starting with all possible turns (shown in Figure 4.6a), the north to west turn is eliminated; after this elimination is made, the three routing algorithms shown in Figure 4.6 can be derived. In Figure 4.6a, the west-first algorithm is shown; in addition to eliminating the North to West turn, the South to West turn is eliminated. In other words, a message must first travel in the West direction before traveling in any other direction. The North-Last algorithm (Figure 4.6b) eliminates both the North to West and the North to East turns. Once a message has turned North, no further turns are permitted; hence, the North turn must be made last. Finally, Figure 4.6c removes turns from North to West and East to South to create the Negative-First algorithm. A message travels in the negative directions (west and south) first before it is permitted to travel in positive directions (east and north). All three of these turn model routing algorithms are deadlock-free. Figure 4.7 illustrates a possible turn elimination that is invalid; the elimination of North to West combined with the elimination of West to North can lead to deadlock. A deadlock cycle is depicted in Figure 4.7b that can result from a set of messages using the turns specified in Figure 4.7a.

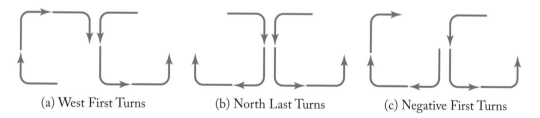

(a) West First Turns   (b) North Last Turns   (c) Negative First Turns

Figure 4.6: Turn model routing.

Odd-even turn model routing [74] proposes eliminating a set of two turns depending on whether the current node is in an odd or even column. For example, when a packet is traversing a node in an even column,[2] turns from East to North and from North to West are prohibited. For packets traversing an odd column node, turns from East to South and from South to West are prohibited. With this set of restrictions, the odd-even turn model is deadlock free provided

---

[2]A column is even if the dimension-0 coordinate of the column is even.

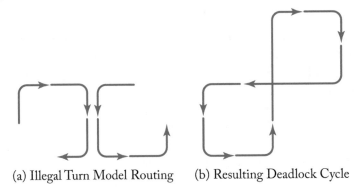

(a) Illegal Turn Model Routing   (b) Resulting Deadlock Cycle

Figure 4.7: Turn model deadlock.

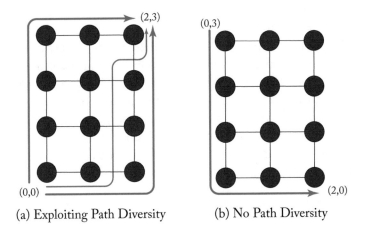

(a) Exploiting Path Diversity   (b) No Path Diversity

Figure 4.8: Negative-first routing example.

180° turns are disallowed. The odd-even turn model provides better adaptivity than other turn model algorithms such as West-First. With West-First, destinations to the West of the source, have no flexibility; with odd-even routing, there is flexibility depending on the allowable turns for a given column.

In Figure 4.8, we apply the Negative-First turn model routing to two different source destination pairs. In Figure 4.8a, three possible routes are shown between (0,0) and (2,3) (more are possible); turns from North to East and from East to North are permitted allowing for significant flexibility. However, in Figure 4.8b, there is only one path allowed by the algorithm to route from (0,3) to (2,0). The routing algorithm does not allow the message to turn from East to South. Negative routes must be completed first, resulting in no path diversity for this source-destination pair. As illustrated by this example, turn model routing provides more flexibility and adaptivity than dimension-order routing but it is still somewhat restrictive.

4.6 MULTICAST ROUTING

So far, we have focused on unicast (i.e., one-to-one) routing algorithms. However, there often occur scenarios where a core needs to send the same message to multiple cores. This is known as a broadcast (if all cores in the system need to be signaled) or a multicast (if a subset of the cores in the system need to be signaled). In shared memory cache coherent systems, examples of this are seen in broadcast-based and in limited directory-based coherence protocols. In message passing systems, this is required by routines like MPI_Bcast. A naïve implementation of multicasts is to simply send multiple unicasts, one per destination. But this increases traffic in the network substantially, leading to poor network latency and throughput [114].

There have been a few proposals to support multicast routing on-chip. Virtual Circuit Tree Multicasting (VCTM) [114] adds small routing tables at every router. For every multicast, one unicast setup packet per destination is sent out before the multicast to configure the routing tables along the XY route. All setup packets for the same multicast destination set carry a unique VCT ID, which corresponds to the index in the routing table. Each setup packet appends its output port to the VCT ID entry in the routing table, thus setting up the directions out of which the multicast flit should get forked. All subsequents multicasts to this destination are injected with this VCT ID and get appropriately forked at the routers in the network. Whirl [206] is a routing algorithm optimized for broadcasts, that tries to create load-balanced broadcast trees on the fly, allowing broadcasts to use different combinations of links, thereby increasing link utilization and throughput. In both of these designs, the router needs to support forking of the same flit out of multiple directions.

4.7 ROUTING ON IRREGULAR TOPOLOGIES

The discussion of routing algorithms in this chapter has assumed a regular topology such as a torus or a mesh. In the previous chapter, the potential for power and performance benefits of using irregular topologies for MPSoCs composed of heterogeneous nodes was explored. Irregular topologies can require special considerations in the development of a routing algorithm. Common routing implementations for irregular networks rely on source table routing or node-table routing [54, 123, 169]. Care must be taken when specifying routes so that deadlock is not induced. Turn model routing may not be feasible if certain connectivity is removed by the presence of oversized cores in a mesh network, for example. Up*/Down* [310] routing is a popular deadlock-free routing algorithm for irregular topologies, that marks each link as either Up or Down, starting from a root node. All flits can only transition from a Up link to a Down link, but never the opposite, which guarantees deadlock freedom.

4.8 IMPLEMENTATION

In this section, we discuss various implementation options for routing algorithms. Routing algorithms can be implemented using look-up tables at either the source nodes or within each router. Combinational circuitry can be used as an alternative to table-based routing. Implementations have various trade-offs, and not all routing algorithms can be achieved with each implementation. Table 4.1 shows examples for how routing algorithms in each of the three different classes can be implemented.

Table 4.1: Routing algorithm and implementation options

Routing Algorithm	Source Routing	Combinational	Node Table
Deterministic			
DOR	Yes	Yes	Yes
Oblivious			
Valiant's	Yes	Yes	Yes
Minimal	Yes	Yes	Yes
Adaptive	No	Yes	Yes

4.8.1 SOURCE ROUTING

Routing algorithms can be implemented in several ways. First, the route can be embedded in the packet header at the source, known as source routing. For instance, the X-Y route in Figure 4.1 can be encoded as $< E, E, N, N, N, Eject >$, while the Y-X route can be encoded as $< N, N, N, E, E, Eject >$. At each hop, the router will read the leftmost direction off the route header, send the packet toward the specified outgoing link, and strip off the portion of the header corresponding to the current hop.

There are a few benefits to source routing. First, by selecting the entire route at the source, latency is saved at each hop in the network since the route does not need to be computed or looked up. The per-router routing hardware is also saved; no combinational routing logic or routing tables are needed once the packet has received its route from the source node. Second, source routing tables can be reconfigured to deal with faults and can support irregular topologies. Multiple routes per source-destination pair can be stored in the table (as shown in Table 4.2) and selected randomly for each packet to improve load balancing.

The disadvantages of source routing include the bit overheads required to store the routing table at the network interface of each source and to store the entire routing path in each packet; these paths are of arbitrary length and can grow large depending on network size. For a 5-port switch, each routing step is encoded by a 3-bit binary number. Just as the packet must be able to handle arbitrary length routing paths, the source table must also be designed to efficiently store

Table 4.2: Source routing table at Node (0,0) for the 2×3 mesh in Figure 4.1

Destination	Route 0	Route 1
00	X	X
10	EX	EX
20	EEX	EEX
01	NX	NX
11	NEX	ENX
21	NEEX	ENEX
02	NNX	NNX
12	ENNX	NENX
22	EENNX	NNEEX
03	NNNX	NNNX
13	NENNX	ENNNX
23	EENNNX	NNNEEX

different length paths. Additionally, by choosing the entire route at the source node, source-based routing is unable to take advantage of dynamic network conditions to avoid congestion. However, as mentioned, multiple routes can be stored in the table and selected either randomly or with a given probability to improve the load distribution in the network.

4.8.2   NODE TABLE-BASED ROUTING

More sophisticated algorithms are realized using routing tables at each hop which store the outgoing link a packet should take to reach a particular destination. By accessing routing information at each hop (rather than all at the source), adaptive algorithms can be implemented and per-hop network congestion information can be leveraged in making decisions.

Table 4.3 shows the routing table for the west-first turn model routing algorithm on a 3-ary 2-mesh. Each node's table would consist of the row corresponding to its node identifier, with up to two possible outgoing links for each destination. Ejection is indicated by an X. By implementing a turn model routing algorithm in the per-node tables, some adaptivity can be achieved.

Compared to source routing, node-based routing requires smaller routing tables at each node. Each routing table needs to store only the routing information to select the next hop for each destination rather than the entire path. When multiple outputs are included per destination, node-based routing supports some adaptivity. Local information about congestion or faults can be used to bias the route selection to the non-congested link or to a non-faulty path. Node-

Table 4.3: Table-based routing for a 3 × 3 mesh with west-first turn model algorithm

From	TO								
	00	01	02	10	11	12	20	21	22
00	X -	N -	N -	E -	E N	E N	E -	N E	N E
01	S -	X -	N -	E S	E -	E N	E S	E -	E N
02	S -	S -	X -	E S	E S	E -	E S	E S	E -
10	W -	W -	W -	X -	N -	N -	E -	E N	E N
11	W -	W -	W -	S -	X -	N -	E S	E -	N E
12	W -	W -	W -	S -	S -	X -	E S	E S	E -
20	W -	W -	W -	W -	W -	W -	X -	N -	N -
21	W -	W -	W -	W -	W -	W -	S -	X -	N -
22	W -	W -	W -	W -	W -	W -	S -	S -	X -

based routing tables can also be programmable. By allowing the routing tables to be changed, the routing algorithm is better able to tolerate faults in the network.

The most significant downside to node routing tables is the increase in packet delay. Source routing requires a single look-up to acquire the entire routing path for a packet. With node-based routing, the latency of a look-up must be expended at each hop in the network.

4.8.3   COMBINATIONAL CIRCUITS

Alternatively, the message can encode the ordinates of the destination and use comparators at each router to determine whether to accept (eject) or forward the message. Simple routing algorithms are typically implemented as combinational circuits within the router due to the low overhead.

With source routing, the packet must contain space to carry all the bits needed to specify the entire path. Routing using combinational circuits requires only that the packet carry the destination identifier. The circuits required to implement the routing algorithm can be quite simple and executed with very low latency. An example circuit to compute the next hop based on current buffer occupancies in a 2-D mesh is shown in Figure 4.9. Alternatively, the route selection could implement dimension order routing rather than make a selection based on queue lengths.

By implementing the routing decision in combinational circuits, the algorithm is specific to one topology and one routing algorithm. The generality and configurability of table-based strategies are sacrificed. Despite the speed and simplicity of using a circuit to compute the next hop in the routing path, this computation adds latency to the packet traversal when compared to source-based routing. As with node-routing, the next output must be determined at each hop

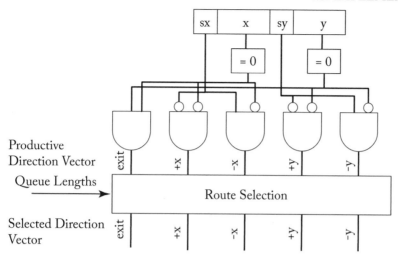

Figure 4.9: Combinational routing circuit for 2-D mesh.

in the network. As will be discussed in Chapter 6, this routing computation can add a pipeline stage to the router traversal.

4.8.4 ADAPTIVE ROUTING

Adaptive routing algorithms need mechanisms to track network congestion levels, and update the route. Route adjustments can be implemented by modifying the header, by employing combinational circuitry that accepts as input these congestion signals, or by updating entries in a routing table. Many congestion sensitive mechanisms have been proposed, with the simplest being tapping into information that is already captured and used by the flow control protocol, such as buffer occupancy or credits [89, 319].

The primary benefit of increasing the information available to the routing circuitry is adaptivity. By improving the routing decision based on network conditions, the network can achieve higher bandwidth and reduce the congestion latency experienced by packets.

The disadvantage of such an approach is complexity. Additional circuitry is required for congestion-based routing decisions; this circuitry can increase the latency of a routing decision and the area of the router. Although the leveraging of information already available at the router is often done to make routing decisions, increasing the sophistication of the routing decision may require that additional information be communicated from adjacent routers. This additional communication could increase the network area and energy.

4.9   BRIEF STATE-OF-THE-ART SURVEY

In this section, we provide a brief overview of current routing algorithm and implementation research in on-chip networks.

Algorithms. Many on-chip network chip prototypes utilize dimension order routing for its simplicity and deadlock freedom (see Chapter 8); however, other routing techniques have been proposed.

Various oblivious routing algorithms [13, 75, 193, 313, 320] have been explored to push bandwidth without the added complexity of adaptive routing. Adaptive routing algorithms have also been investigated [74, 125, 137, 190, 233, 301, 319], with various papers focusing on the implementation overhead of adaptive routing given the tight design constraints of on-chip networks [137, 233, 236, 301]. Routing algorithms that dynamically switch between adaptive and deterministic have been proposed [162]. Ætheral employs source-based routing and relies on the turn model for deadlock freedom [135]. Flattened butterfly [186] is an example of an on-chip network that uses non-minimal routing to improve load balancing in the network; non-minimal routes are also employed in a bufferless on-chip network to prevent packets from being dropped [252]. Deflective routing [262] attempts to route packets to a free virtual channel along the minimal path but will misroute when this is not possible. Customized routing can be specified for application-specific designs with well understood communication patterns [163, 257].

This chapter has focused most of its discussion of routing algorithms on unicast routing, that is routing a packet from a single source to a single destination. Recent research has also explored the need to support for routing of collective communication including multicast routing [2, 114, 206, 234, 304, 354] and many-to-one routing [206, 234].

Implementation. Various table-based implementations of routing algorithms have been explored for on-chip networks. Node table-based routing is proposed for regular topologies [124, 304], while table-based routing for irregular topologies have also been explored [54, 123, 169, 278]. Multicast routing has been proposed using a table-based implementation [114, 304] or a circuit-based implementation [206, 234, 354].

Fault-tolerant Routing. There has been a recent interest in designing routing algorithms for on-chip networks where certain links might fail due to soft/hard errors. These designs add routing tables into each router and reconfigure these upon fault detection with deadlock-free routes [20, 216].

CHAPTER 5

Flow Control

Flow control governs the allocation of network buffers and links. It determines when buffers and links are assigned to messages, the granularity at which they are allocated, and how these resources are shared among the many messages using the network. A good flow control protocol lowers the latency experienced by messages at low loads by not imposing high overhead in resource allocation, and drives up network throughput by enabling effective sharing of buffers and links across messages. In determining the rate at which packets access buffers (or skip buffer access altogether) and traverse links, flow control is instrumental in determining network energy and power consumption. The implementation complexity of a flow control protocol includes the complexity of the router microarchitecture as well as the wiring overhead required for communicating resource information between routers.

5.1   MESSAGES, PACKETS, FLITS, AND PHITS

When a message is injected into the network, it is first segmented into packets, which are then divided into fixed-length flits, short for flow control units. For instance, a 128-byte cache line sent from a sharer to a requester will be injected as a message, and if the maximum packet size is larger than 128 bytes, the entire message will be encoded as a single packet. The packet will consist of a head flit that contains the destination address, body flits, and a tail flit that indicates the end of a packet. Flits can be further broken down into phits, which are physical units and correspond to the physical channel width. The breakdown of messages to packets and packets to flits is depicted in Figure 5.1a. Head, body, and tail flits all contain parts of the cache line and the cache coherence command. Each flit also contains certain control information such as flit type and virtual channel number. For instance, if flit size is 128 bits, the 128-byte packet will consist of 8 flits: 1 head, 6 body, and 1 tail, ignoring the extra bits needed to encode the destination and other information needed by the flow control protocol. In short, a message is the logical unit of communication above the network, and a packet is the physical unit that makes sense to the network. A packet contains destination information while a flit may not, thus all flits of a packet must take the same route.

   Due to the abundance of on-chip wiring resources, channels tend to be wider in on-chip networks, so messages are likely to consist of a single packet. In off-chip networks, channel widths are limited by pin bandwidth; this limitation causes flits to be broken down into smaller chunks called phits. To date, in on-chip networks, flits are composed of a single phit and are the smallest subdivision of a message due to wide on-chip channels. Additionally, as illustrated in

(margin notes) message · packet · flit · head, body and tail flits · phits

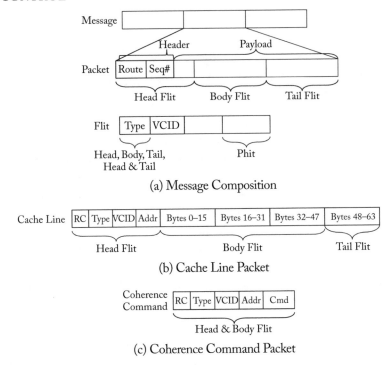

(a) Message Composition

(b) Cache Line Packet

(c) Coherence Command Packet

Figure 5.1: Composition of message, packets, flits: Assuming 16-byte wide flits and 64-byte cache lines, a cache line packet will be composed of 5 flits and a coherence command will be a single-flit packet. The sequence number (Seq#) is used to match incoming replies with outstanding requests, or to ensure ordering and detect lost packets.

Figure 5.1, many messages will in fact be single-flit packets. For example, a coherence command need only carry the command and the memory address which can fit in a 16-byte wide flit.

Flow control techniques are classified by the granularity at which resource allocation occurs. We will discuss techniques that operate on message, packet and flit granularities in the next sections with a table summarizing the granularity of each technique at the end.

5.2 MESSAGE-BASED FLOW CONTROL

We start with circuit-switching, a technique that operates at the message level, which is the coarsest granularity, and then refine these techniques to finer granularities.

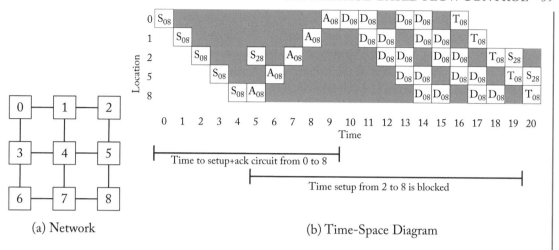

(a) Network   (b) Time-Space Diagram

Figure 5.2: Circuit-switching example from Core 0 to Core 8, with Core 2 being stalled. S: Setup flit, A: Acknowledgement flit, D: Data message, T: Tail (deallocation) flit. Each D represents a message; multiple messages can be sent on a single circuit before it is deallocated. In cycles 12 and 16, the source node has no data to send.

5.2.1   CIRCUIT SWITCHING

Circuit switching pre-allocates resources (links) across multiple hops to the entire message. A probe (a small setup message) is sent into the network and reserves the links needed to transmit the entire message (or multiple messages) from the source to the destination. Once the probe reaches the destination (having successfully allocated the necessary links), the destination transmits an acknowledgement message back to the source. When the source receives the acknowledgement message, it releases the message which can then travel quickly through the network. Once the message completes its traversal, the resources are deallocated. After the setup phase, per-hop latency to acquire resources is avoided. With sufficiently large messages, this latency reduction can amortize the cost of the original setup phase. In addition to possible latency benefits, circuit switching is also bufferless. As links are pre-reserved, buffers are not needed at each hop to hold packets that are waiting for allocation, thus saving power. While latency can be reduced, circuit switching suffers from poor bandwidth utilization. The links are idle between setup and the actual message transfer and other messages seeking to use those resources are blocked.

Figure 5.2 illustrates an example of how circuit-switching flow control works. Dimension order X-Y routing is assumed with the network shown in Figure 5.2a. As time proceeds from left to right (Figure 5.2b), the setup flit, S constructs a circuit from Core 0 to Core 8 by traversing the selected route through the network. At time 4, the setup flit has reached the destination and begins sending an acknowledgement flit, A back to Core 0. At time 5, Core 2 wants to initiate a transfer to Core 8; however, the resources (links) necessary to reach Core 8 are already allocated to Core 0. Therefore, Core 2's request is stalled. At time 9, the acknowledgment request is received by Core 0 and the data transfers, D can begin. Once the required data are sent, a tail flit, T is sent by Core 0 to deallocate these resources. At time 19, Core 2 can now begin acquiring the resources recently deallocated by the tail flit. From this example, we see that there is significant wasted link bandwidth. During the setup time and when links have been reserved but there is no data that needs to be transmitted these links are idle but unavailable to other messages (wasted link bandwidth is shaded in grey). Core 2 also suffers significant latency waiting for resources that are mostly idle.

Asynchronous transfer mode (ATM) [102] establishes virtual circuit connections; before data can be sent, network resources must be reserved from source to destination (like circuit switching). However, data are switched through the network at a packet granularity rather than a message granularity.

5.3   PACKET-BASED FLOW CONTROL

Circuit-switching allocates resources to messages and does so across multiple network hops. There are several inefficiencies to this scheme; next, we look at schemes that allocate resources to packets. Packet-based flow control techniques first break down messages into packets, then interleave these packets on the links, thus improving link utilization. Unlike circuit switching, the remaining techniques will require per-node buffering to store in-flight packets.

5.3.1   STORE AND FORWARD

With packet-based techniques, messages are broken down into multiple packets and each packet is handled independently by the network. In *store-and-forward* flow control [86], each node waits until an entire packet has been received before forwarding any part of the packet to the next node. As a result, long delays are incurred at each hop, which makes them unsuitable for on-chip networks that are usually delay-critical. Moreover, store and forward flow control requires that there be sufficient buffering at each router to buffer the entire packet. These high buffering requirements reduce store and forward switching's amenability to on-chip networks.

In Figure 5.3, we depict a packet traveling from Core 0 to Core 8 using store and forward switching. Once the tail flit has been buffered at each node, the head can then allocate the next link and depart for the next router. Serialization delay is paid for at each hop for the body and tail flits to catch up with the head flit. For a 5-flit packet, the latency is 5 cycles to transmit the packet at each hop.

5.3.2   VIRTUAL CUT-THROUGH

To reduce the delay packets experience at each hop, *virtual cut-through* flow control [180] allows transmission of a packet to proceed to the next node before the entire packet is received at the

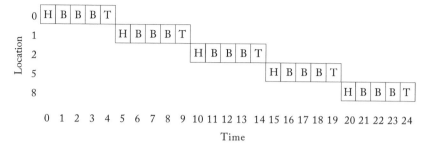

Figure 5.3: Store and forward example.

current router. Latency experienced by a packet is thus drastically reduced over store and forward flow control, as shown in Figure 5.4a. In Figure 5.3, 25 cycles are required to transmit the entire packet; with virtual cut-through, this delay is reduced to 9 cycles. However, bandwidth and storage are still allocated in packet-sized units. Packets still move forward only if there is enough storage at the next downstream router to hold the entire packet. On-chip networks with tight area and power constraints may find it difficult to accommodate the large buffers needed to support virtual cut-through when packet sizes are large (such as 64- or 128-byte cache lines).

In Figure 5.4b, the entire packet is delayed when traveling from node 2 to node 5 even though node 5 has buffers available for 2 out of 5 flits. No flits can proceed until all 5 flit buffers are available.

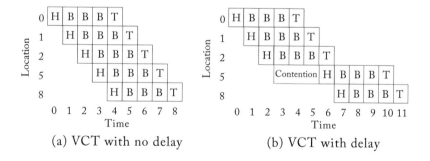

(a) VCT with no delay   (b) VCT with delay

Figure 5.4: Virtual cut through example.

5.4   FLIT-BASED FLOW CONTROL

To reduce the buffering requirements of packet-based techniques, flit-based flow control mechanisms exist. Low buffering requirements help routers meet tight area or power constraints on-chip.

5.4.1 WORMHOLE

Like virtual cut-through flow control, *wormhole* flow control [93] cuts through flits, allowing flits to move on to the next router before the entire packet is received at the current location. For wormhole flow control, the flit can depart the current node as soon as there is sufficient buffering for this flit. However, unlike store-and-forward and virtual cut-through flow control, wormhole flow control allocates storage and bandwidth to flits rather than entire packets. This allows relatively small flit buffers to be used in each router, even for large packet sizes. While wormhole flow control uses buffers effectively, it makes inefficient use of link bandwidth. Though it allocates storage and bandwidth in flit-sized units, a link is held for the duration of a packet's lifetime in the router. As a result, when a packet is blocked, all of the physical links held by that packet are left idle. Since wormhole flow control allocates buffers on a flit granularity, a packet composed of many flits can potentially span several routers, which will result in many idle physical links. Throughput suffers because other packets queued behind the blocked packet are unable to use the idle physical links.

In the example in Figure 5.5, each router has 2 flit buffers. When the head flit experiences contention traveling from 1–2, the remaining two body and tail flits are stalled at Core 0 since there is no buffer space available at Core 1 until the head moves to Core 2. However, the channel is still held by the packet even though it is idle as shown in grey.

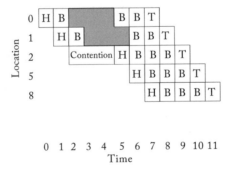

Figure 5.5: Wormhole example.

Wormhole flow control reduces packet latency by allowing a flit to leave the router as soon as a downstream buffer is available (in the absence of contention, the latency is the same as virtual cut through). Additionally, wormhole flow control can be implemented with fewer buffers than packet-based techniques. Due to the tight area and power constraints of on-chip networks, wormhole flow control is the predominant technique adopted thus far.

5.5   VIRTUAL CHANNELS

Virtual channels have been explained as the "swiss-army knife" of interconnection networks [86]. They were first proposed as a solution for deadlock avoidance [87], but have also been applied to mitigate head-of-line blocking in flow control, thus extending throughput. Head-of-line blocking occurs in all the above flow control techniques where there is a single queue at each input; when a packet at the head of the queue is blocked, it stalls subsequent packets that are lined up behind it, even when there are available resources for the stalled packets.

Essentially, a virtual channel (VC) is basically a separate queue in the router; multiple VCs share the physical wires (physical link) between two routers. By associating multiple separate queues with each input port, head-of-line blocking can be reduced. Virtual channels arbitrate for physical link bandwidth on a cycle-by-cycle basis. When a packet holding a virtual channel becomes blocked, other packets can still traverse the physical link through other virtual channels. Thus, VCs increase the utilization of the physical links and extend overall network throughput.

Technically, VCs can be applied to all the above flow control techniques to alleviate head-of-line blocking, though Dally first proposed them with wormhole flow control [87]. For instance, circuit switching can be applied on virtual channels rather than the physical channel, so a message reserves a series of VCs rather than physical links, and the VCs are time-multiplexed onto the physical link cycle-by-cycle, also called virtual circuit switching [128]. Store-and-forward flow control can also be used with VCs, with multiple packet buffer queues, one per VC, VCs multiplexed on the link packet-by-packet. Virtual cut-through flow control with VCs work similarly, except that VCs are multiplexed on the link flit-by-flit. However, as on-chip network designs overwhelmingly adopt wormhole flow control for its small area and power footprint, and use virtual channels to extend the bandwidth where needed, for the rest of this book, when we mention virtual channel flow control, we assume that it is applied to wormhole, with both buffers and links managed and multiplexed at the granularity of flits.

A walk-through example illustrating the operation of virtual channel flow control is depicted in Figure 5.6. Packet A initially occupies VC 0 and is destined for Node 4, while Packet B initially occupies VC 1 and is destined for Node 2. At time 0, Packet A and Packet B both have flits waiting in the west input virtual channels of Node 0. Both A and B want to travel outbound on the east output physical channel. The head flit of Packet A is allocated virtual channel 0 for the west input of router 1 and wins switch allocation (techniques to handle this allocation are discussed in Chapter 6). The head flit of packet A travels to router 1 at time 1. At time 2, the head flit of packet B is granted switch allocation and travels to router 1 and is stored in virtual channel 1. Also at time 2, the head flit of A fails to receive a virtual channel for router 4 (its next hop); both virtual channels are occupied by flits of other packets. The first body flit of A inherits virtual channel 0 and travels to router 1 at time 3. Also at time 3, the head flit of B is able to allocate virtual channel 0 at router 2 and continues on. At time 4, the first body flit of packet B inherits virtual channel 1 from the head flit and wins switch allocation to continue to router 1. By time 7, all of the flits of B have arrived at router 2, the head and body flits have continued

head-of-line blocking

virtual channel

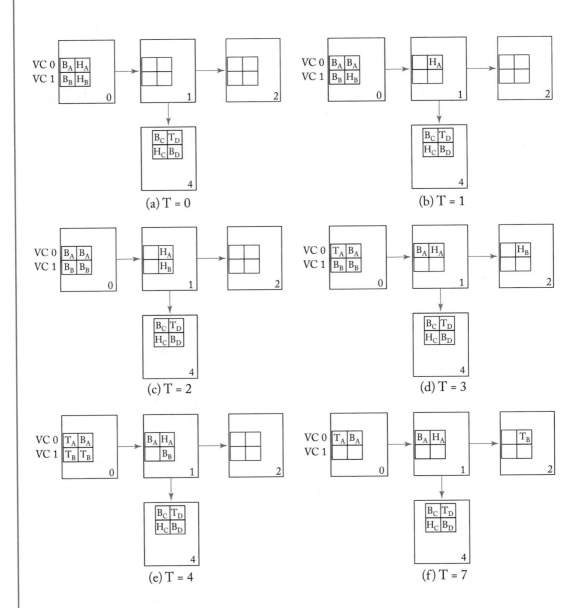

Figure 5.6: Virtual channel flow control walk-through example. Two packets A and B are broken into 4 flits each (H: head, B: Body, T: Tail).

on and the tail flit remains to be routed. The head flit of packet A is still blocked waiting for a free virtual channel to travel to router 4.

With wormhole flow control using a single virtual channel, packet B would be blocked behind packet A at router 1 and would not be able to continue to router 2 despite the availability of buffers, links and the switch to do so. Virtual channels allow packet B to proceed toward its destination despite the blocking of packet A. Virtual channels are allocated once at each router to the head flit and the remainder of flits inherit that virtual channel. With virtual-channel flow control, flits of different packets can be interleaved on the same physical channel, as seen in the example between time 0 and 2.

Virtual channels are also widely used to break deadlocks, both within the network (see Section 5.6), and for handling system-level or protocol-level deadlocks (see Section 2.1.3).

The previous sections have explained how different techniques handle resource allocation and utilization. These techniques are summarized in Table 5.1.

Table 5.1: Summary of flow control techniques

	Links	Buffers	Comments
Circuit-Switching	Messages	N/A (buffer-less)	Requires setup and acknowledgment
Store and Forward	Packet	Packet	Head flit must wait for entire packet before proceeding on next link
Virtual Cut Through	Packet	Packet	Head can begin next link traversal before tail arrives at current node
Wormhole	Packet	Flit	Head of line blocking reduces efficiency of link bandwidth
Virtual Channel	Flit	Flit	Can interleave flits of different packets on links

5.6   DEADLOCK-FREE FLOW CONTROL

Deadlock freedom can be maintained either through the use of constrained routing algorithms that ensure no cycles ever occur (see Chapter 4), or through the use of deadlock-free flow control which allows any routing algorithm to be used.

5.6.1   DATELINE AND VC PARTITIONING

Figure 5.7 illustrates how two virtual channels can be used to break a cyclic deadlock in the network when the routing protocol permits a cycle. Here, since each VC is associated with a separate buffer queue, and every VC is time-multiplexed onto the physical link cycle-by-cycle, holding onto a VC implies holding onto its associated buffer queue rather than locking down a physical link. By enforcing an order on these VCs, so that lower-priority VCs cannot request and wait for higher-priority VCs, there can be no cycle in resource usage. In Figure 5.7, all messages are sent through VC 0 until they cross the dateline. After crossing the dateline, messages are assigned to VC 1 and cannot be allocated to VC 0 at any point during the remainder of their network traversal [94]. This ensures that the channel dependency graph (CDG) [94] is acyclic.

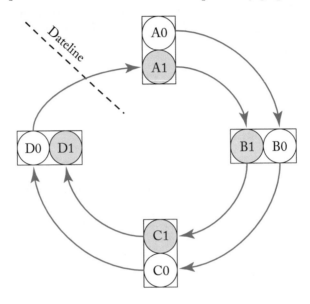

Figure 5.7: Two virtual channels with separate buffer queues denote with white and grey circles at each router are used to break the cyclic route deadlock in Figure 4.2.

The same idea works across various oblivious/adaptive routing algorithms that allow all turns and are thus deadlock-prone. A routing algorithm that randomly chooses between X-Y and Y-X routes can be made deadlock-free by enforcing all X-Y packets to use VC 0 and all Y-X packets to use VC 1. Similarly, routing algorithms that wish to allow all turns for path diversity can be made deadlock-free by implementing a certain turn model in VC 0 and another turn model in VC 1, and not allowing packets in one VC to jump to the other throughout the traversal.

At the system level, messages that can potentially block each other can be assigned to different message classes that are mapped to different virtual channels within the network, such as request and acknowledgment messages of coherence protocols. These designs scale to multiple VCs by dividing all available VCs into multiple classes, and enforcing the ordering rules described above across these classes. Within each class, flits can acquire any VC. Implementation complexity of virtual channel routers will be discussed in detail next in Chapter 6 on router microarchitecture.

5.6.2   ESCAPE VCS

The previous section discussed the benefits of enforcing ordering between VCs to prevent dead-locks. However, enforcing an order on VCs lowers their utilization, affecting network through-put when the number of VCs is small. In Figure 5.7, all packets are initially assigned VC 0 and remain on VC 0 until they cross the dateline. As a result, VC 1 is underutilized. Escape VCs have been proposed to address this by Duato [108]. Duato proved that the requirement of an acyclic CDG was a sufficient condition for a deadlock-free routing algorithm but not necessary; even if the CDG is cyclic, as long as there is an acyclic sub-part of the CDG, it can be used to escape out of the cyclic-dependency. This acyclic connected sub-part of the CDG defines a escape virtual channel. Hence, rather than enforcing a fixed order/priority between all VCs, that so long as there is a single escape VC that is deadlock-free, all other VCs can use fully adaptive routing with no routing restrictions. This escape VC is typically made deadlock-free by using a deadlock-free routing function within it. For instance, if VC 0 is designated as the escape channel, all traffic on VC 0 must be routed using dimension-ordered routing, while all other VCs can be routed with arbitrary routing functions. Explained simply, so long as access to VCs is arbitrated fairly, a packet always has a chance of landing on the escape VC, and thus of escaping a deadlock.[1] Escape VCs help increase the utilization of VCs, or permits a higher throughput with a smaller number of VC, making for leaner routers.

escape VC

In Figure 5.8a, we illustrate once again how unrestricted routing with a single virtual channel can lead to deadlock. Each packet is trying to acquire resources to make a clockwise turn. Figure 5.8b utilizes two virtual channels. Virtual channel 1 serves as an escape virtual channel. For example, Packet A could be allocated virtual channel 1 (and thus dimension order routed to its destination). By allocating virtual channel 1 at router 4 for packet A, all packets can make forward progress. The flits of packet A will eventually drain from VC 0 at router 1, allowing packet B to be allocated either virtual channel at router 1. Once the flits of packet B have drained, packet D can continue on virtual channel 0 or be allocated to virtual channel 1 and make progress before packet B has drained. The same goes for the flits of packet C.

5.6.3   BUBBLE FLOW CONTROL

An alternate idea to avoid deadlocks, without requiring multiple VC classes, is to ensure that a closed cyclic-dependency between buffers is never created at runtime. k-ary, n-cubes are dead-lock prone due to the presence of a ring network in each dimension. This ring network inherently produces a cyclic dependence even when deadlock-free routing such as DOR is used. Bubble Flow Control [298] is used in combination with virtual cut-through to provide deadlock free-dom in k-ary, n-cube networks. Packets currently traveling within a particular dimension are handled as normal by virtual cut through flow control. Packets needing to be injected into the network or change dimensions are handled based on bubble flow control which controls the

---

[1]If minimal routing is used, a packet is allowed to hop in and out of an escape VC. But for non-minimal routes, a packet that gets into an escape VC has to continue in it to provide deadlock-freedom.

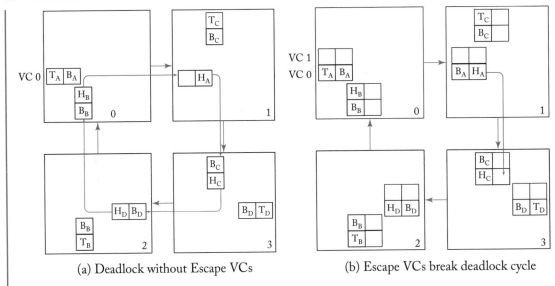

Figure 5.8: Escape virtual channel example. Virtual Channel 1 serves as an escape virtual channel that is dimension order XY routed.

injection into the ring to make sure a closed cyclic dependency is not created. A packet can only be injected if there is empty buffer space in the ring to accommodate two packets. Requiring empty buffer space for two packets guarantees that if the packet is injected, there will still be one empty packet buffer in the ring. This empty buffer, referred to as a *bubble*, ensures that at least one packet in the ring will be able to make forward progress, thus preventing the cycle to close. Figure 5.9 shows an example where R1 has two empty bubbles which will allow Packet P1 to be injected. The remaining routers only have one free bubble each preventing the injection of Packets P0 and P2. The rule same applies for packets changing dimensions which is considered as injection into a new dimension.

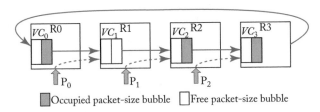

Figure 5.9: Bubble flow control example.

Due to the complexity associated with searching all buffers in a ring, bubble flow control requires there be two empty packet buffers in the local queue in order for a packet to be injected [298]. This increases the minimum buffer sizes requirements which can be undesirable for maintaining a low area and power footprint in on-chip networks. Recent work has explored adapting bubble flow control to wormhole switching to reduce the buffering requirements and make it more compatible with on-chip networks [68, 147, 237, 355].

5.7 BUFFER BACKPRESSURE

As most on-chip network designs cannot tolerate the dropping of packets, there must be buffer backpressure mechanisms for stalling flits. Flits must not be transmitted when the next hop will not have buffers available to house them. The unit of buffer backpressure depends on the specific flow control protocol; store-and-forward and virtual cut-through flow control techniques manage buffers in units of packets, while wormhole and virtual channel flow control manage buffers in units of flits. Circuit switching, being a bufferless flow control technique, does not require buffer backpressure mechanisms. Two commonly used buffer backpressure mechanisms are credits and on/off signaling.

Credit-based. Credits keep track of the number of buffers available at the next hop, by sending a credit to the previous hop when a buffer is vacated (when a flit/packet leaves the router), and incrementing the credit count at the previous hop upon receiving the credit. When a flit departs the current router, the current router decrements the credit count for the appropriate downstream buffer.

credit backpressure

On/off. On/off signaling involves a signal between adjacent routers that is turned off to stop the previous hop from transmitting flits when the number of buffers drop below a threshold. This threshold must be set to ensure that all in-flight flits will have buffers upon arrival. Buffers must be available for flits departing the current router during the transmission latency of the off-signal. When the number of free buffers at the downstream router rises above a threshold, the signal is turned on and flit transmission can resume. The on threshold should be selected so that the next router will still have flits to send to cover the time of transmission of the on signal plus the delay to receive a new flit from the current router.

on-off backpressure

5.8 IMPLEMENTATION

The implementation complexity of a flow control protocol essentially involves the complexity of the entire router microarchitecture and the wiring overhead imposed in communicating resource information between routers. Here, we focus on the latter, as Chapter 6 elaborates on router microarchitectures and associated implementation issues.

When choosing a specific buffer backpressure mechanism, we need to consider its performance in terms of buffer turnaround time, and its overhead in terms of the number of reverse signaling wires.

5.8.1   BUFFER SIZING FOR TURNAROUND TIME

buffer turnaround
time

Buffer turnaround time is the minimum idle time between when successive flits can reuse a buffer. A long buffer turnaround time leads to inefficient reuse of buffers, which results in poor network throughput. If the number of buffers implemented does not cover the buffer turnaround time, then the network will be artificially throttled at each router, since flits will not be able to flow continuously to the next router even when there is no contention from other ports of the router. As shown in Figure 5.10, the link between two routers is idle for 6 cycles while waiting for a free buffer at the downstream router.

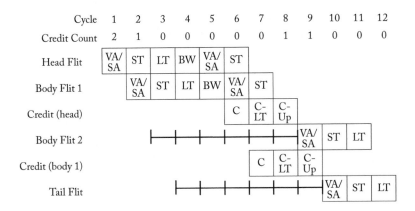

Figure 5.10: Throttling due to too few buffers. Flit pipeline stages discussed in Chapter 6. C: Credit send. C-LT: Credit link traversal. C-Up: Credit update.

For credit-based buffer backpressure, a buffer is held from the time a flit departs the current node (when the credit counter is decremented), to the time the credit is returned to inform the current node that the buffer has been released (so the credit counter can be incremented again). Only then can the buffer be allocated to the next flit, although it is not actually reused until the flit traverses the current router pipeline and is transmitted to the downstream router. Hence, the turnaround time of a buffer is at least the sum of the propagation delay of a data flit to the next node, the credit delay back, and the pipeline delay, as is shown in Figure 5.11a.

In comparison, in on/off buffer backpressure, a buffer is held from the time a flit arrives at the next node and occupies the last buffer (above the threshold), triggering the off signal to be sent to stop the current node from sending. This persists until a flit leaves the next node and frees up a buffer (causing the free buffer count to go over the threshold). Consequently, the on signal is asserted, informing the current node that it can now resume sending flits. This buffer is occupied again when the data flit arrives at the next node. Here, the buffer turnaround time is thus at least twice the on/off signal propagation delay plus the propagation delay of a data flit, and the pipeline delay, as shown in Figure 5.11b.

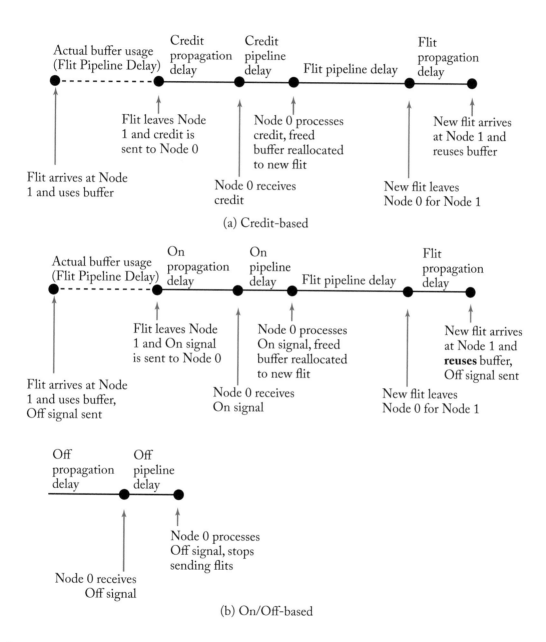

Figure 5.11: Buffer backpressure mechanisms time lines.

If we have a 1-cycle propagation delay for both data flits and reverse signaling between adjacent nodes, a 1-cycle pipeline delay for buffer backpressure signals, and a 3-cycle router pipeline, then credit-based backpressure will have a buffer turnaround time of at least 6 cycles, while on/off backpressure will have a buffer turnaround time of at least 8 cycles. Note that this implies that this network using on/off backpressure needs at least 8 buffers per port to cover the turnaround time, while if it chooses credit-based backpressure, it needs 2 fewer buffers per port. Thus, buffer turnaround time also affects the area overhead, since buffers take up a substantial portion of a router's footprint.

Note that it is possible to optimize the buffer turnaround time by triggering the backpressure signals (credits or on/off) once it is certain a flit will depart a router and no longer need its buffer, rather than waiting until the flit has actually been read out of the buffer.

5.8.2   REVERSE SIGNALING WIRES

While on/off backpressure performs poorly compared to credit-based backpressure, it has lower overhead in terms of reverse signaling overhead. Figure 5.12 illustrates the number of reverse signaling wires needed for both backpressure mechanisms: credit-based requires log_B wires per queue (virtual channel), where B is the number of buffers in the queue, to encode the credit count. On the other hand, on/off needs only a single wire per queue. With eight buffers per queue and two virtual channels, credit-based backpressure requires six reverse signaling wires (Figure 5.12a) while on/off requires just two reverse wires (Figure 5.12b).

In on-chip networks, where there is abundant on-chip wiring, reverse signaling overhead tends to be less of a concern than area overhead and throughput. Hence, credit-based backpressure will be more suitable.

5.9   FLOW CONTROL IN APPLICATION SPECIFIC ON-CHIP NETWORKS

time division
multiplexing
(TDM)

Multiprocessor SoCs (MPSoCs) typically rely on wormhole flow control for the same reasons that it has been adopted for more general purpose on-chip networks. Applications that run on MPSoCs often have real-time performance requirements. Such quality of service requirements can impact flow control design decisions. The network interface controller can regulate traffic injected into the network to reduce contention and ensure fairness [262, 343]. The use of time division multiplexing (TDM) that allocates a fixed amount of bandwidth to each node is one way to provide guaranteed throughput and avoid contention [135]. Time division multiplexing schedules communications so that each communication flow has their own time slot on network links. The size and number of time slots implemented governs the granularity at which network resources can be allocated. As the TDM slots are allocated a priori, deviation in actual bandwidth demands will lead to idling channels.

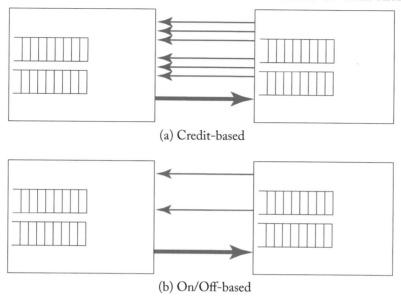

(a) Credit-based

(b) On/Off-based

Figure 5.12: Reverse signaling overhead.

A custom network for an MPSoC may result in a heterogeneous set of switches; these switches may differ in terms of number of ports, number of virtual channels, and number of buffers [47]. For the flow control implementation, different numbers of buffers may be instantiated at each node depending on the communication characteristics [164]. Buffering resources will impact the thresholds of on/off flow control or the reverse signaling wires required by credit-based flow control. Additionally, non-uniform link lengths in a customized topology will impact the buffer turn-around time of the flow control implementation. Regularity and modularity are sacrificed in this type of environment; however, the power, performance, and area gains can be significant.

5.10   BRIEF STATE-OF-THE-ART SURVEY

Wormhole flow control is widely used in many on-chip network prototype chips, such as MIT RAW [335], Tilera TILE64 [356], UT Austin TRIPS operand network (OPN) [138, 139], and Princeton Piton [39]. All of these designs use multiple physical networks to boost bandwidth. Layered switching [231] hybridizes wormhole and virtual cut-through flow control by allocating resources to groups of data words which are larger than flits but smaller than packets. Virtual Channels have also been used across on-chip network proposals and prototype multicore chips, such as Intel TeraFLOPS [158], Intel SCC [159], and MIT SCORPIO [101].

Other research into flow control for on-chip networks has targeted the unique opportunities and constraints in on-chip networks, not afforded by off-chip networks. For instance, several works leverage the availability of upper metal layers and heterogeneous interconnect sizing to harness links of different speeds and bandwidths in flow control [37, 290]. Others target the tight power constraint faced by on-chip networks, and propose flow control protocols that allow flits to bypass through router pipelines, lowering dynamic switching power at routers, while improving buffer turnaround time and latency-throughput [209, 210], or dynamically scale down the number of VCs needed to support the traffic demands [260]. SMART proposes a flow control mechanism to allow multiple hops in the network to be traversed in a single cycle [204]. Novel flow control techniques based on bubble flow control make efficient use of a small number of buffers to improve throughput without paying area and power penalties [68, 235, 237]. The theory behind bubble flow control—a bubble in a dependence ring can guarantee forward progress—has been leveraged and extended recently to provide deadlock recovery in any dependence ring that gets created at runtime in any irregular topology due to faults or dynamic power-gating [302].

The overheads of fault tolerance for various flow control strategies have been explored [299]. Elastic buffer flow control [243] uses pipelined channels for buffering, eliminating the need for virtual channel buffers. To avoid the area and power overheads of buffers, several bufferless flow control techniques have been explored on chip [117, 148, 252]. A flow control scheme that can adapt between bufferless and buffered routing to achieve the performance and energy advantages of both has been proposed [168]. Several proposals aim to achieve the benefits of both circuit and packet switching [18, 113, 357, 360].

Most commonly in on-chip networks, phits are the same size as flits; however, some work explores using smaller flits than the physical channel width and develops mechanisms to allow flits from multiple packets to share the channel simultaneously [353]. Decoupling flit and phit width has also been explored to facilitate dynamic reconfiguration of channel bandwidth [75, 154, 214]

Flow control techniques have also been leveraged to provide quality of service [142, 143, 275]. Different packets can be given priority to access network resources such as buffers and links to provide low latency or higher throughput for more critical packets. Recent QoS schemes distinguish between latency sensitive and insensitive packets within a single application [99, 317] and across multiple applications [98].

CHAPTER 6

Router Microarchitecture

Routers must be designed to meet latency and throughput requirements amid tight area and power constraints; this is a primary challenge designers are facing as many-core systems scale. Router complexity increases with bandwidth demands; very simple routers (unpipelined, wormhole, no VCs, limited buffering) with low area and power overheads can be built when high throughput is not needed. Challenges arise when the latency and throughput demands on on-chip networks are raised.

A router's microarchitecture determines its critical path delay which affects per-hop delay and overall network latency. The implementation of the routing, flow control, and the actual router pipeline affect the efficiency at which buffers and links are used which governs overall network throughput. Router microarchitecture also impacts network energy—both dynamic and leakage—as it determines the circuit components in a router and their activity. Finally, the microarchitecture and underlying circuits directly contribute to the area footprint of the network.

6.1 VIRTUAL CHANNEL ROUTER MICROARCHITECTURE

Figure 6.1 shows the microarchitecture of a state-of-the-art credit-based virtual channel (VC) router to explain how typical routers work. The example assumes a 2-D mesh, so the router has five input and output ports corresponding to the four neighboring directions and the local processing element (PE) port. The major components which constitute the router are the input buffers, route computation logic, virtual channel allocator, switch allocator, and the crossbar switch. Most on-chip network routers are input-buffered, in which packets are stored in buffers only at the input ports, as input buffering permits the use of single-ported memories. Here, we assumed four VCs at each input port, each with its own buffer queue that is four flits deep.

The buffers are responsible for storing flits when they enter the router, and housing them throughout their duration in the router. This is in contrast to a processor pipeline that latches instructions in buffers between each pipeline stage. If source routing is not used, the route computation block will compute (or lookup) the correct output port for this packet. The allocators (virtual channel and switch) determine which flits are selected to proceed to the next stage where they traverse the crossbar. Finally, the crossbar switch is responsible for physically moving flits from the input port to the output port.

Over the next few sections, we discuss the various components inside a router.

input buffered

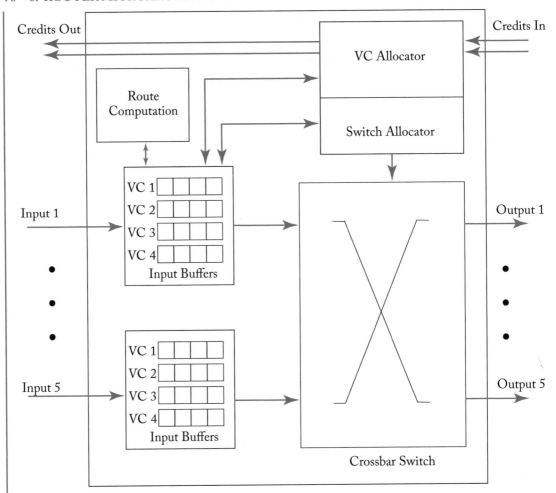

Figure 6.1: A credit-based virtual channel router microarchitecture.

6.2   BUFFERS AND VIRTUAL CHANNELS

Buffers are used to house packets or flits when they cannot be forwarded right away onto output links. Flits can be buffered on the input ports and on the output ports. Output buffering occurs when the allocation rate of the switch is greater than the rate of the channel. Crossbar speedup (discussed in Section 6.3.2) requires output buffering since multiple flits can be allocated to a single output channel in the same cycle.

All previously proposed on-chip network routers have buffering at input ports, as input buffer organization permits area and power-efficient single-ported memories. We will, therefore, focus our discussion on input-buffered routers here, dissecting how such buffering is organized *within* each input port.

6.2.1 BUFFER ORGANIZATION

Buffer organization has a large impact on network throughput, as it heavily influences how efficiently packets share link bandwidth.

Single fixed-length queue. Figure 6.2a shows an input-buffered router where there is a single queue in each input port, i.e., there are no VCs. Incoming flits are written into the tail of the queue, while the flit at the head of the queue is read and sent through the crossbar switch and onto the output links (when it wins arbitration). The single queue has a fixed length, so the upstream router can keep track of buffer availability and ensure that a flit is forwarded only if there is a free buffer downstream.

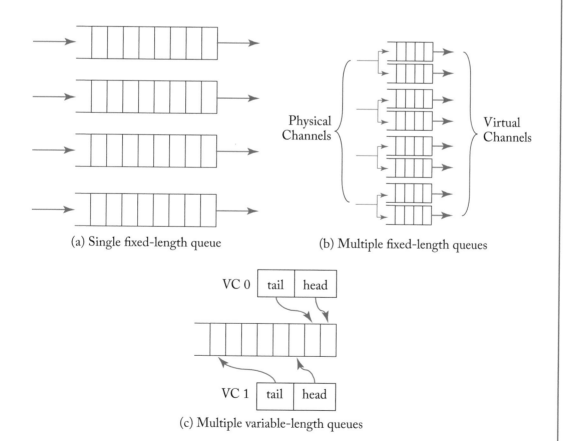

(a) Single fixed-length queue

(b) Multiple fixed-length queues

(c) Multiple variable-length queues

Figure 6.2: Buffer and VC organizations.

Clearly, a single queue can lead to scenarios where a packet at the head of the queue is blocked (as its output port is held by another packet), while a packet further behind in the queue whose output port is available could not make forward progress as it has to wait for the head of the queue to clear. Such unnecessary blocking is termed head-of-line blocking.

Multiple fixed-length queues. Having multiple queues at each input port helps alleviate head-of-line blocking. Each of these queues is termed a virtual channel, with multiple virtual channels multiplexing and sharing the physical channel/link bandwidth. Figure 6.2b shows an input-buffered router where there are two separate queues in each input port, corresponding to a router with 2 VCs.

Multiple variable-length queues. In the above buffer organization, each VC has a fixed-length queue, sized at four flits in Figure 6.2b. If there is imbalance in the traffic, there could be a VC that is full and unable to accept more flits when another VC is empty, leading to poor buffer utilization and thus low network throughput.

To get around this, each VC queue can be variable-length, sharing a large buffer [334], as shown in Figure 6.2c. This permits better buffer utilization, but at the expense of more complex circuitry for keeping track of the head and tail of the queues. Also, to avoid deadlocks, one flit buffer needs to be reserved for each VC, so that other VCs will not fill up the entire shared buffer and starve out a VC, ensuring forward progress.

Minimum number of virtual channels. VCs serve two purposes in NoCs—deadlock (protocol or routing) avoidance and performance improvement. For the former, a certain number of VCs would be required to avoid protocol deadlock (for instance requests vs. responses in shared memory coherence protocols). These VCs are often known as virtual networks. Within each virtual network, additional VCs may be required to serve as escape VCs to avoid routing deadlock, as described earlier in Chapter 5.

Apart from these required VCs, additional VCs can be added to improve performance by removing/mitigating head-of-line blocking. With the same total amount of buffering per port, designers have the choice of using many shallow VCs or fewer VCs with deeper buffers. More VCs further ease head-of-line blocking and thus improve throughput. It, however, comes at the expense of a more complex VC allocator and VC buffer management. Furthermore, the efficiency of many, shallow VCs vs. few, deep VCs will depend on the traffic pattern. With light traffic, many shallow VCs will lead to under utilization of extra VCs. Under periods of heavy traffic, few, deep VCs will be less efficient as packets will be blocked due to a lack of available VCs.

Minimum number of buffers. For functional correctness, a router needs at least one buffer per virtual channel to avoid deadlocks. This is because packets in two different VCs should never indefinitely block one another. Beyond that, for sustaining full throughput, there needs to be a minimum number of buffers (within each VC or in total, depending on the buffer organization) to cover the buffer turnaround time, which was discussed in Chapter 5.

6.2.2 INPUT VC STATE

Each Virtual Channel is associated with the following state for flits sitting in it.

Global (G): Idle/Routing/waiting for output VC/waiting for credits in output VC/Active. Active VCs can perform switch allocation.

Route (R): Output port for the packet. This field is used for switch allocation. The output port is populated after route computation by the head flit. In designs with lookahead routing (described later in Section 6.5.2) or source routing, the head flit arrives at the current router with the output port already designated.

Output VC (O): Output VC (i.e., VC at downstream router) for this packet. This is populated after VC allocation by the head flit, and used by all subsequent flits in the packet.

Credit Count (C): Number of credits (i.e., flit buffers at downstream router) in output VC O at output port R. This field is used by body and tail flits.

Pointers (P): Pointers to head and tail flits. This is required if buffers are implemented as a shared pool of multiple variable-length queues, as described above.

6.3 SWITCH DESIGN

The crossbar switch of a router is the heart of the router datapath. It switches bits from input ports to output ports, performing the essence of a router's function.

6.3.1 CROSSBAR DESIGNS

Aggressive design of crossbar switches at high frequencies and low power is a challenge in VLSI design, such as the bit-interleaved or double-pumped custom crossbar used in the Intel TeraFLOPs chip [158]. Here, we just provide some background on basic crossbar designs, and discuss alternative microarchitectural organizations of crossbars.

Table 6.1 shows a Verilog module describing a crossbar switch, where input select signals to each multiplexer set up the connections of the switch, i.e., which input port(s) should be connected to which output port(s). Synthesizing this will lead to a crossbar composed of many multiplexers, such as that illustrated in Figure 6.3. Most low-frequency router designs will use such synthesized crossbars.

As designs push toward GHz clock and are faced with more stringent power budgets, custom-designed crossbars tend to be used [158, 287, 315]. These have crosspoint-based organizations with select signals feeding each crosspoint, setting up the connection of the switch, like that in Figure 6.4.

With either design, a switch's area and power scale at $O((pw)^2)$, where p is the number of crossbar ports and w is the crossbar port width in bits. A router architect thus has to diligently choose p, a function of the topology, and w, which affects flit size and thus overall packet energy-delay.

multiplexer crossbar

crosspoint crossbar

Table 6.1: Verilog of a 4-bit 5-port crossbar

```
module xbar (clk, reset, in0, in1, in2, in3, in4, out0, out1, out2
             out3, out4, colsel0, colsel1,colsel2, colsel3, colsel4);
input clk;
input reset;
input['CHANNELWIDTH:0] in0, in1, in2, in3, in4;
output['CHANNELWIDTH:0] out0, out1, out2, out3, out4;
input [2:0] colsel0, colsel1, colsel2, colsel3 colsel4;
reg [2:0] colsel0reg, colsel1reg, colsel2reg, colsel3reg, colsel4reg;

bitxbar bx0(in0[0],in1[0],in2[0],in3[0],in4[0],out0[0],out1[0],out2[0] out3[0],
            out4[0],colsel0reg,colsel1reg,colsel2reg,colsel3reg,colsel4reg,1'bx);

bitxbar bx1(in0[1],in1[1],in2[1],in3[1],in4[1],out0[1],out1[1],out2[1],out3[1],
            out4[1],colsel0reg,colsel1reg,colsel2reg,colsel3reg,colsel4reg,1'bx);

bitxbar bx2(in0[2],in1[2],in2[2],in3[2],in4[2],out0[2],out1[2],out2[2],out3[2],
            out4[2],colsel0reg,colsel1reg,colsel2reg,colsel3reg,colsel4reg,1'bx);

bitxbar bx3(in0[3],in1[3],in2[3],in3[3],in4[3],out0[3],out1[3],out2[3],out3[3]
            out4[3],colsel0reg,colsel1reg,colsel2reg,colsel3reg,colsel4reg,1'bx);

endmodule
module bitxbar(i0,i1,i2,i3,i4,o0,o1,o2,o3,o4,sel0,sel1,sel2,sel3,sel4,inv
input i0,i1,i2,i3,i4;
output o0,o1,o2,o3,o4;
[2:0] sel0, sel1, sel2, sel3, sel4;
input inv;

buf b0(i00, i0); //buffer for driving in0 to the 5 muxes
...
buf b4(i40, i4);

mux5_1 m0(i00, i10, i20, i30, i40, o0, sel0, inv);
...
mux5_1 m4(i00, i10, i20, i30, i40, o4, sel4, inv);

endmodule
```

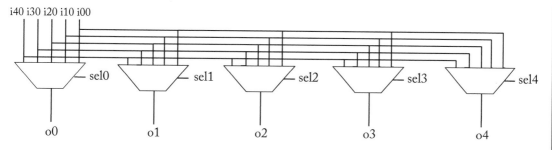

Figure 6.3: Crossbar composed of many multiplexers.

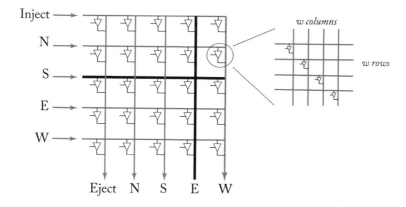

Figure 6.4: A 5×5 crosspoint crossbar switch. Each horizontal and vertical line is w bits wide (1 phit width). The bold lines show a connection activated from the south input port to the east output port.

6.3.2 CROSSBAR SPEEDUP

A router microarchitect needs to decide on the crossbar switch speedup, i.e., the number of input and output ports in the crossbar relative to the number of router input and output ports. Figure 6.5 shows various alternative crossbar designs with different speedup factors: crossbars with higher speedups provide more internal bandwidth between router input and output ports, and thus ease the allocation problem and improving flow control. For instance, if each VC has its own input port to the crossbar, a flit can be read out of every VC every cycle, so multiple VCs need not contend for the same crossbar input port. A 10×5 crossbar (such as shown in Figure 6.5b) will achieve close to 100% throughput even with a simple allocator (allocators are discussed in the next section). By providing more inputs to select from, there is a higher probability that each output port will be matched (used) each cycle. The use of output speedup allows multiple flits to be sent to the same output port each cycle, thus reducing the contention. A crossbar with output speedup requires output buffers to multiplex flits onto single output port.

input speedup

output speedup

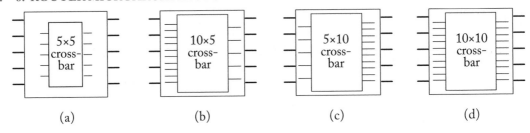

Figure 6.5: Crossbars with different speedups for a 5-port router. (a) No crossbar speedup, (b) crossbar with input speedup of 2, (c) crossbar with output speedup of 2, and (d) crossbar with input and output speedup of 2.

Crossbar speedup can also be achieved by clocking the crossbar at a higher frequency than the rest of the router. For instance, if the crossbar is clocked at twice the router frequency, it can then send two flits each cycle between a single pair of input-output ports, achieving the same performance as a crossbar with input and output speedup of 2. This is less likely in on-chip networks where a router tends to run off a single clock supply that is already aggressive.

6.3.3   CROSSBAR SLICING

With the crossbar taking up a significant portion of a router's footprint and power budget, microarchitectural techniques targeted toward optimizing crossbar power-performance have been proposed.

dimension slicing

Dimension slicing a crossbar [265] in a 2-D mesh uses two 3×3 crossbars instead of one 5×5 crossbar, with the first crossbar for traffic that remains in the X-dimension, and the second crossbar for traffic remaining in the Y-dimension. A port on the first crossbar connects with a port on the second crossbar so traffic that turns from the X to Y dimension traverses both crossbars while those remaining within a dimension traverses only one crossbar. This is particularly suitable for the dimension-ordered routing protocol where traffic mostly stays within a dimension.

bit interleaving

Bit interleaving the crossbar targets w instead. It sends alternate bits of a link on the two phases of a clock on the same line, thus halving w. The TeraFLOPS architecture employs bit interleaving, as will be discussed in Chapter 8.

6.4   ALLOCATORS AND ARBITERS

allocator: N to M

arbiter: N to 1

switch arbiter

An allocator matches N requests to M resources while an arbiter matches N requests to 1 resource. In a router, the resources are VCs (for virtual channel routers) and crossbar switch ports.

In a wormhole router with no VCs, the switch arbiter at each output port matches and grants that output port to requesting input ports. Hence, there are P arbiters, one per output port, each arbiter matching P input port requests to the single output port under contention.

In a router with multiple VCs, we need a virtual-channel allocator (VA), which resolves contention for output virtual channels and grants them to input virtual channels, as well as a switch allocator (SA) that grants crossbar switch ports to input virtual channels. Only the head flit of a packet needs to access the virtual-channel allocator, while the switch allocator is accessed by all flits and grants access to the switch on a cycle-by-cycle basis.

An allocator/arbiter that delivers high matching probability translates to more packets succeeding in obtaining virtual channels and passage through the crossbar switch, thereby leading to higher network throughput. In most NoCs, the allocation logic in the router determines cycle time. Thus allocators and arbiters must be fast and pipeline-able so they can work under high clock frequencies.

VC allocator

switch allocator

6.4.1   ROUND-ROBIN ARBITER

With a round-robin arbiter, the last request to be serviced will have the lowest priority in the next round of arbitration. Figure 6.6 shows the circuit required for a round-robin arbiter. If Grant$_i$ is high, Priority$_{i+1}$ becomes high in the next cycle and all other priorities become low.

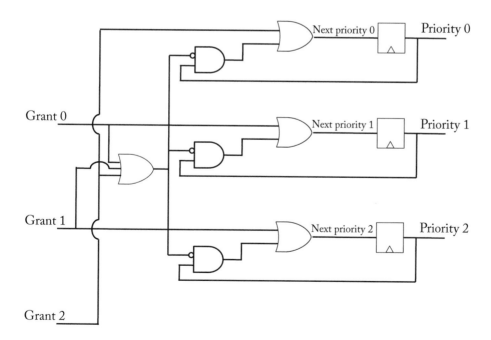

Figure 6.6: Round-robin arbiter.

Next, we will walk through an example granting requests with a round-robin arbiter. A set of requests from 4 different requestors are shown in Figure 6.7. Suppose the last request serviced prior to this set of requests was from Requestor A. As a result, B has the highest priority at the start of the example. With the round-robin arbiter, requests are satisfied in the following order: $B_1, C_1, D_1, A_1, D_2, A_2$.

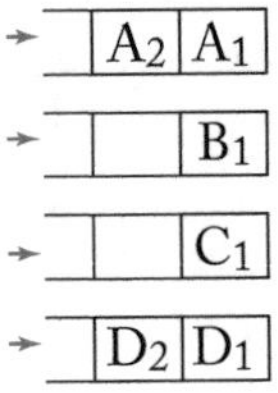

Figure 6.7: Request queues for arbiter examples.

6.4.2   MATRIX ARBITER

A matrix arbiter operates so that the least recently served requestor has the highest priority [86]. The implementation of a matrix arbiter is shown in Figure 6.8. A triangular array of state bits w_{ij} are stored to implement priorities, with $w_{ij} = \neg w_{ji} \forall i \neq j$. When bit w_{ij} is set, request i has a higher priority than request j. When a request line is asserted, the request is AND-ed with the state bits in that row to disable any lower priority requests. Each time a request k is granted, the state of the matrix is updated by clearing all bits in row k and setting all bits in column k.

Next, we will walk-through the same set of requests from the previous example (Figure 6.7). The initial state of the matrix arbiter is given in Figure 6.9a. Since there are 4 requesters A, B, C, and D, the matrix is 4×4. While only the upper triangle values of the matrix need to be stored, all values are shown for clarity. As each request is granted, the updated matrix values are shown. We can see from the initial state of the matrix that requestor D has the highest priority, followed by C, followed by B, and followed by A. This is because bits [1,0], [2,0], [3,0], [2,1], [3,1], and [3,2] are all set to 1. At T=1, D_1 is granted. As a result, the bits in the 4th row are cleared and bits in the 4th column are set (Figure 6.9b). C now has the highest priority. At T=2, request C_1 is granted and the 3rd row cleared and the 3rd column set, resulting in the matrix in Figure 6.9c. Now B has the highest priority. Grants continue in this fashion with the resulting grant order being $D_1, C_1, B_1, A_1, D_2, A_2$.

6.4.3   SEPARABLE ALLOCATOR

To reduce the complexity allocators and make them pipeline-able, allocators can be built as a composition of multiple arbiters. Recall that arbiters choose one out of multiple requests to a single resource. For instance, a $N : M$ allocator can be built by using N/k arbiters, each $k : 1$ in the first stage to get k candidates from the N initial requestors, followed by M arbiters, each

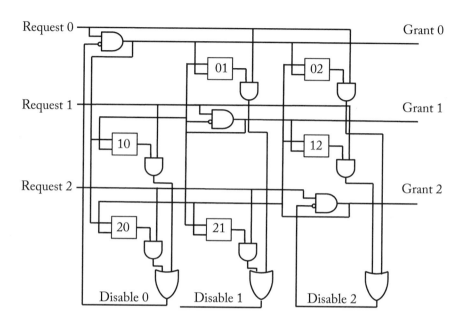

Figure 6.8: Matrix arbiter. The boxes w_{ij} represent priority bits. When bit w_{ij} is set, request i has a higher priority than request j.

$$
\begin{pmatrix}
x & 0 & 0 & 0 \\
1 & x & 0 & 0 \\
1 & 1 & x & 0 \\
1 & 1 & 1 & x
\end{pmatrix}
\begin{pmatrix}
x & 0 & 0 & 1 \\
1 & x & 0 & 1 \\
1 & 1 & x & 1 \\
0 & 0 & 0 & x
\end{pmatrix}
\begin{pmatrix}
x & 0 & 1 & 1 \\
1 & x & 1 & 1 \\
0 & 0 & x & 0 \\
0 & 0 & 1 & x
\end{pmatrix}
\begin{pmatrix}
x & 1 & 1 & 1 \\
0 & x & 0 & 0 \\
0 & 1 & x & 0 \\
0 & 1 & 1 & x
\end{pmatrix}
$$

$\quad$ (a) T = 0 $\qquad\qquad$ (b) T = 1 $\qquad\qquad$ (c) T = 2 $\qquad\qquad$ (d) T = 3

$$
\begin{pmatrix}
x & 0 & 0 & 0 \\
1 & x & 0 & 0 \\
1 & 1 & x & 0 \\
1 & 1 & 1 & x
\end{pmatrix}
\begin{pmatrix}
x & 0 & 0 & 1 \\
1 & x & 0 & 1 \\
1 & 1 & x & 1 \\
0 & 0 & 0 & x
\end{pmatrix}
\begin{pmatrix}
x & 0 & 0 & 0 \\
1 & x & 0 & 1 \\
1 & 1 & x & 1 \\
1 & 0 & 0 & x
\end{pmatrix}
$$

$\qquad$ (e) T = 4 $\qquad\qquad$ (f) T = 5 $\qquad\qquad$ (g) T = 6

Figure 6.9: Matrix arbiter priority update for the request stream from Figure 6.7.

$N/k : 1$ to generate M grants. k can be some parameter specific to the design. Figure 6.10 shows an example; here a 3:4 separable allocator (an allocator matching 3 requests to 4 resources) is composed of arbiters. For instance, consider a separable switch allocator for a router with four ports, and three input VCs per input port. During the first stage of the allocator (comprised of four 3:1 arbiters), each arbiter corresponds to an input port and chooses one of the three input VCs as a winner. The winning VCs from the first stage then arbitrate for an output port in the second stage (comprising three 4:1 arbiters). Each arbiter chooses one out of these input VCs as a winner for the output port. Different arbiters have been used in practice, with round-robin arbiters being the most popular due to their simplicity.

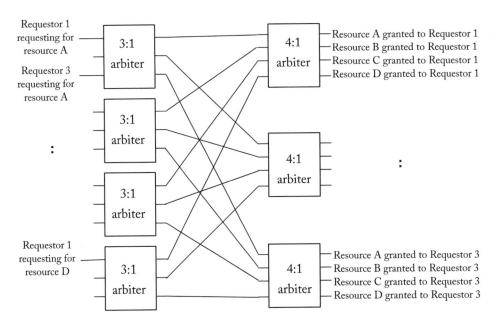

Figure 6.10: A separable 3:4 allocator (3 requestors, 4 resources) which consists of four 3:1 arbiters in the first stage and three 4:1 arbiters in the second. The 3:1 arbiters in the first stage decides which of the 3 requestors win a specific resource, while the 4:1 arbiters in the second stage ensure a requestor is granted just 1 of the 4 resources.

Figure 6.11 shows one potential outcome from a separable allocator. Figure 6.11a shows the request matrix. Each of the 3:1 arbiters selects one value of each row of the matrix; these first stage results of the allocator are shown in the matrix in Figure 6.11b. The second set of 4:1 arbiters will arbitrate among the requests set in the intermediate matrix. The final result (Figure 6.11c) shows that only one of the initial requests was granted. Depending on the arbiters used and the initial states, more allocations could result.

$$\begin{pmatrix} 1 & 1 & 1 \\ 1 & 1 & 0 \\ 1 & 0 & 0 \\ 1 & 0 & 1 \end{pmatrix} \qquad \begin{pmatrix} 1 & 0 & 0 \\ 1 & 0 & 0 \\ 1 & 0 & 0 \\ 0 & 0 & 0 \end{pmatrix} \qquad \begin{pmatrix} 1 & 0 & 0 \\ 0 & 0 & 0 \\ 0 & 0 & 0 \\ 0 & 0 & 0 \end{pmatrix}$$

(a) Request Matrix   (b) Intermediate Matrix   (c) Grant Matrix

Figure 6.11: Separable allocator example.

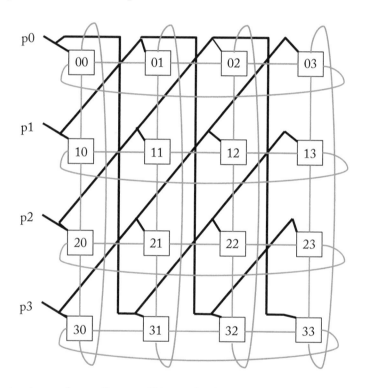

Figure 6.12: A 4 × 4 wavefront allocator. Diagonal priority groups are connected with bold lines. Connections for passing tokens are shown with grey lines.

6.4.4   WAVEFRONT ALLOCATOR

The challenge with separable allocators is often an inefficiency in matching requests to resources, since the first stage is oblivious of the outcome of the second stage. A wavefront allocator performs the entire allocation as one step and is much more efficient, while being implementable in hardware. Figure 6.12 shows a 4 × 4 wavefront allocator [333] which is used in the SGI SPI-DER chip [126] and the Intel SCC [159]. Non-square allocators can be realized by adding dummy rows or columns to create a square array. The 3×4 allocation example shown above with a separable allocator requires a 4×4 wavefront allocator.

The execution of a wavefront allocator begins with setting one of the four priority lines (p0...p3). This supplies row and column tokens to the diagonal group of cells connected to the selected priority line. If one of the cells is requesting a resource, it will consume the row and column tokens and its resource request is granted. Cells that cannot use their tokens pass row tokens to the right and column tokens down. To improve fairness, the initial priority group changes each cycle.

Using the same request matrix from Figure 6.11a, we next illustrate the function of a wavefront allocator. Shaded in light grey are the requests from the request matrix. The first diagonal wave of priorities starting with p0 is circled in Figure 6.13a. Entry [0,0] is the first to receive a grant (highlighted in dark grey). Next, the wave propagates down and to the right (shown in Figure 6.13b). Entry [0,0] consumed a token in the first stage when it received its grant; therefore, as the wave propagates, [0,1] and [1,0] do not receive a token in the second wave since it was already consumed. Entry [3,2] receives tokens from [3,1] and from [2,2] which results in its request being granted (Figure 6.13c). Figure 6.13c shows the 3rd priority wave; the remaining unused tokens are again passed down and to the right. Request [1,1] receives valid tokens in this wave and receives a grant.

After the wavefronts have fully propagated, the grant matrix that results is shown in Figure 6.14. The wavefront allocator is able to grant three requests (as opposed to the single request for this example with a separable allocator).

6.4.5 ALLOCATOR ORGANIZATION

Adaptive routing can complicate the switch allocation for flits. For a deterministic routing algorithm, there is a single desired output port; the switch allocator's function is simply to bid for the single output port. With an adaptive routing function that returns multiple candidate output ports, the switch allocator can bid for all output ports. The granted output port must match the virtual channel granted by the virtual channel allocator. Alternatively, the routing function can return a single candidate output port and then retry routing (for a different output port) if the flit fails to obtain an output virtual channel.

The design of the virtual-channel allocator can depends on the implementation of the routing function. The routing function can be implemented to return a single virtual channel. This would lead to a virtual channel allocator that needs to arbitrate only between input virtual channels contending for the same output virtual channel. If the routing implementation is more general and returns multiple candidate virtual channels for the same physical channel, the allocator needs to first arbitrate among v possible first stage requests before forwarding the winning requests to the second stage (can be done with the separable allocator described above). A routing function that returns all candidate virtual channels for all candidate physical channels is the most general and requires more functionality from the virtual-channel allocator.

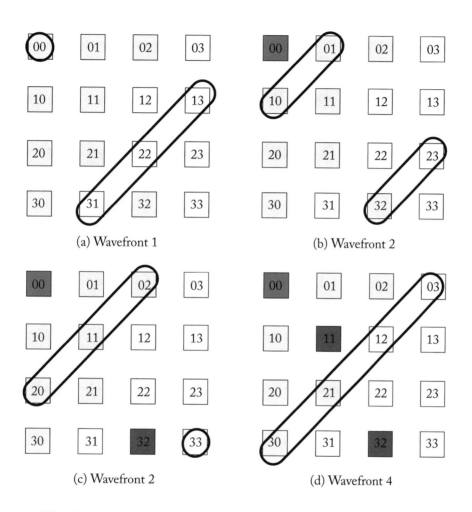

Figure 6.13: Wavefront allocator example.

$$\begin{pmatrix} 1 & 0 & 0 \\ 0 & 1 & 0 \\ 0 & 0 & 0 \\ 0 & 0 & 1 \end{pmatrix}$$

Figure 6.14: Wavefront grant matrix.

With a speculative virtual channel router, non-speculative switch requests must have a higher priority than speculative requests. One way to achieve this is to have two parallel switch allocators. One allocator handles non-speculative requests, while the second handles speculative requests. With the output of both allocators, successful non-speculative requests can be selected over speculative ones. However, if there are no non-speculative requests in the router, then a speculative switch request will succeed in allocating the desired output port. It is possible for a flit to succeed in speculative switch allocation but fail in the parallel virtual channel allocation. In this case the speculation is incorrect and the crossbar passage that was reserved by the switch allocator is wasted. Only head flits are required to perform VC allocation. As a result, subsequent body and tail flits are marked as non-speculative (for their switch allocation) since they inherit the VC allocated to the head flit.

6.5   PIPELINE

Figure 6.15a shows the logical pipeline stages for a basic virtual channel router, with all the components discussed so far. Like the logical pipeline stages of a typical processor: instruction fetch, decode, execute, memory and writeback, these are logical stages that will fit into a physical pipeline depending on the actual clock frequency.

buffer write
route
computation

VC allocation

switch allocation

switch traversal

link traversal

A head flit, upon arriving at an input port, is first decoded and buffered according to its input VC in the buffer write (BW) pipeline stage. Next, the routing logic performs route computation (RC) to determine the output port for the packet. The header then arbitrates for a VC corresponding to its output port (i.e., the VC at the next router's input port) in the VC allocation (VA) stage. Upon successful allocation of a VC, the header flit proceeds to the switch allocation (SA) stage where it arbitrates for the switch input and output ports. On winning the output port, the flit is then read from the buffer and proceeds to the switch traversal (ST) stage, where it traverses the crossbar. Finally, the flit is passed to the next node in the link traversal (LT) stage. Body and tail flits follow a similar pipeline except that they do not go through RC and VA stages, instead inheriting the route and the VC allocated by the head flit. The tail flit, on leaving the router, deallocates the VC reserved by the head flit.

A wormhole router with no VCs does away with the VA stage, requiring just four logical stages. In Figure 6.1, such a router will not require a VC allocator, and will have only a single deep buffer queue in each input port.

6.5.1   PIPELINE IMPLEMENTATION

The logical virtual channel pipeline consists of five stages. A router that is running at a low clock frequency will be able to fit all five stages into a single clock cycle. For aggressive clock frequencies, the router architecture must be pipelined. The actual physical pipeline depends on the implementation of each of these logical stages and their critical path delay in that technology. We discuss implementations of each of these stages later in this chapter.

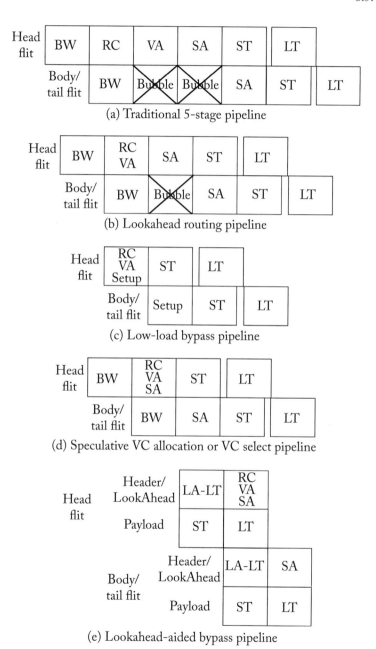

Figure 6.15: Router pipeline [BW: Buffer Write, RC: Route Computation, VA: Virtual Channel Allocation, SA: Switch Allocation, ST: Switch Traversal, LT: Link Traversal].

If the physical pipeline has five stages just like the logical stages, then the stage with the longest critical path delay will set the clock frequency. Typically, this is the VC or Switch allocation stage when the number of VCs is high, or the crossbar traversal stage with very wide, highly ported crossbars. The clock frequency can also be determined by the overall system clock, for instance sized by the processor pipeline's critical path instead.

Increasing the number of physical pipeline stages increases the per-hop router delay for each message, as well as the buffer turnaround time which affects the minimum buffering needed and affects throughput. Thus, pipeline optimizations have been proposed and employed to reduce the number of stages. Common optimizations targeting logical pipeline stages are explained next. State-of-the-art router implementations can perform all actions within a single cycle.

6.5.2   PIPELINE OPTIMIZATIONS

The collective goal of all routers is to enable multiple flows to multiplex over shared resources (links and buffers). Myriad pipeline optimizations have been proposed for on-chip routers to help run various logical stages in parallel, shaving off cycles from some/all routers. This in turn saves latency and energy. Shallow pipelines also lower buffer turnaround time, helping improve network throughput.

lookahead routing

Lookahead Routing [126] removes the RC stage from the critical path. The route of the packet is determined one hop in advance and encoded within the head flit, enabling incoming flits to compete for VCs/switch immediately after the BW stage. The route computation for the next hop can be performed in parallel with VC/switch allocation since it is no longer needed to determine which output ports to arbitrate for. Figure 6.15b shows the router pipeline with lookahead routing, which is also known as next route compute (NRC) or route pre-computation.

low-load bypass

Low-load bypassing removes the BW and SA stages from routers that are lightly loaded. Incoming flits are allowed to speculatively enter the ST stage if there are no flits ahead of it in the input buffer queue. Figure 6.15c shows the pipeline where a flit goes through a single stage of switch setup, during which the crossbar is set up for flit traversal in the next cycle while simultaneously allocating a free VC corresponding to the desired output port, followed by ST and LT. Upon a output port conflict however, the flit is written into the buffer (BW) and subsequently performs SA.

Figure 6.16a shows an example of low-load bypass. At Time 1, A arrives at the South input port and there are no buffered flits waiting in the input queue. The lookahead routing computation is performed in the first cycle (1a) and the crossbar connection between the south input and the east output is setup (1b). At Time 2, A traverses the crossbar and exits the router. Buffering and allocation are bypassed. In Figure 6.16b, two flits arrive at Time 1 (A on the South input and B on the North input); both have empty input queues and attempt to bypass the pipeline. However, during the crossbar setup (1b), a port conflict is detected as both flits attempt to setup the crossbar for the East output port. Now, both flits must be written into input buffers (1c) and go through the regular pipeline.

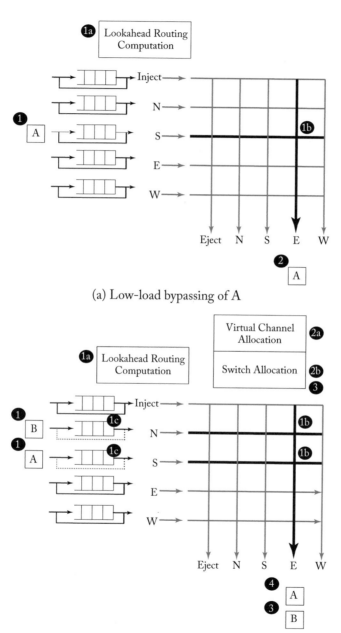

(a) Low-load bypassing of A

(b) Low-load bypassing aborted due to contention between A and B

Figure 6.16: Low-load bypass example.

Speculative VA [254, 255, 292] removes the VA stage from the critical path. A flit enters the SA stage *speculatively* after BW and arbitrates for the switch port while at the same time trying to acquire a free VC. If the speculation succeeds, the flit directly enters the ST pipeline stage. However, when speculation fails, the flit must go through some of these pipeline stages again, depending on where the speculation failed. Figure 6.15d shows the router pipeline with the speculative pipeline (BASE + LA-RC + BY + S-VA).

In Figure 6.16b, during step 2, At time 2, both flits A and B perform virtual channel and switch allocation in parallel (2a and 2b). Packet B successfully allocates an output virtual channel and the switch and traverses the switch at Time 3. Packet A succeeds in virtual channel allocation but fails in switch allocation. At Time 3, A will again attempt to allocate the switch. A's request is now non-speculative since it already obtained an output virtual channel (2a). A's request is successful and it traverses the switch and exits the router at Time 4.

VC selection [101, 208, 287] eliminates the VA stage from the router pipeline. The idea behind VC selection is that a full-fledged VA for multiple output VCs is unnecessary for the buffered flits since only one flit can go out of an output port in any cycle. A queue of free VC ids is maintained at every output port. The SA winner at each output port is assigned the VCid at the head of the queue. A head flit enters SA only if the free VC queue at its output port is non-empty (i.e., the input port at the next router has at least one free VC). Body and tail flits can enter SA without this check. The update of the free VC queue occurs off the critical path. If there are multiple message classes/virtual networks, free VC queues have to be maintained per virtual network, and the SA stage might be stretched to accommodate the extra mux to choose between the heads of each queue. The pipeline is the same as the speculative VC one (Figure 6.15d) except that there is no speculation involved.

Lookahead bypass [101, 208, 209, 210, 287] leverages the above optimization to design a single-cycle router. It removes the BW and SA stages from the critical path of the flit traversal. The idea is to perform SA for a flit at the *next* router while the flit is traversing the link between the current and next router. This is implemented by sending a few bits in advance, called *lookaheads* to the next router while the flit is in ST. These lookaheads are nothing but the header information of the flit (route, VCid, etc.) allowing the channel bandwidth to simply be re-apportioned without requiring extra wires. While the flit performs LT, its lookahead performs SA at the next router. Successful arbitration by lookaheads allows its flit to bypass BW and SA and directly go into ST, reducing its delay to two cycles at every hop (ST+LT). This is shown in Figure 6.15e. This methodology not only saves latency but also buffer read/write power. If the lookahead arbitration fails, the flit gets buffered and goes through the normal pipeline. Flits that get buffered would pay the power cost anyway and just pay an additional latency penalty if there is a competing lookahead or flit which gets higher priority.

Figure 6.17 shows an example where two lookaheads arrive at the router in Cycle 1. B's lookahead wins switch allocation and selects a VC; flit B bypasses buffering and directly enters

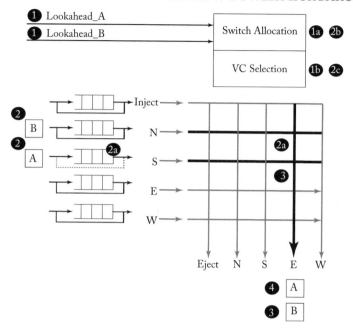

Figure 6.17: Lookahead bypass example—Lookahead_B wins and B bypasses, A gets buffered.

the switch in Cycle 2. Since A's lookahead loses, flit A gets buffered; A performs switch allocation and VC selection in Cycle 2 and performs switch traversal in Cycle 3.

State-of-the-art networks can be designed today at modern technologies that spend a single-cycle for switch arbitration and VC selection in the router, and the subsequent cycle for traversing both the switch and link [160, 287], while operating at GHz frequencies. This enables two-cycles per-hop traversal (at no contention). | State-of-the-art

6.6   LOW-POWER MICROARCHITECTURE

Power consumption has been a challenge since the 1990s for both embedded and high-performance chips. Since the mid-2000s, it has become the primary constraint in most designs. Multicores were an answer to the power problem, and the resulting communication substrate, namely the on-chip network, plays an active role in contributing to the total power consumption of multicores today—both dynamic and leakage.

Figure 6.18a plots the power distribution for a state-of-the-art mesh router with four VCs. These numbers are from chip measurements at 32 nm [64]. At low-loads, the dynamic power component of the buffers and other state (VCs and credits) is primarily due to the clocked latches, rather than the traffic itself. At saturation (i.e., high-loads), buffers contribute 55% of the dynamic power, while the crossbar and links contribute 34%. Static power contributes to over 75% of the total power consumption within the router at low-loads, and 53% at high loads.

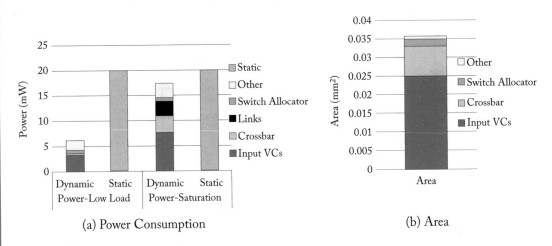

(a) Power Consumption   (b) Area

Figure 6.18: Power and area of a 1-cycle mesh router at 32 nm [64].

In this section, we discuss the techniques used across on-chip networks to reduce power consumption. We refer readers to the *Synthesis Lectures on Computer Architecture Techniques for Power Efficiency* [179] for a more detailed description of low-power techniques used in cores and caches.

6.6.1   DYNAMIC POWER

The equation for dynamic power consumption is $P = \alpha C V^2 f$, where α is the activity factor, C is the capacitance being switched, V is the operating voltage, and f is the operating frequency. To reduce power consumption, there are two classes of techniques. The first tries to reduce power consumption by dynamically reducing V and f, while the second tries to reduce α and C.

DVFS. Dynamic voltage and frequency scaling (DVFS) is the most popular design technique to reduce power consumption of digital circuits. DVFS can be applied to on-chip networks by leveraging the idea that a router with less traffic can be made to operate at lower voltage-frequency state without affecting the aggregate performance. Two key challenges with using DVFS for on-chip network fabric are as follows.

(1) For multiple voltage-frequency islands, bi-synchronous FIFOs have to be used at the interfaces of every pair of different voltage-frequency islands, incurring excess delays.

(2) Most existing proposals assume the use of multiple supply lines for accessing different voltages. However, use of multiple voltage rails requires multiple voltage converters outside the chip along with the area overhead for multiple power distribution networks. The

introduction of high bandwidth integrated voltage regulators can alleviate this problem by allowing fast (sub 50 ns) voltage transitions.

As the on-chip network associated with a tile/core not only serves the flits injected from that core, but also serves flits from different cores, the DVFS policy of the on-chip network fabric has to be dealt with differently than for the cores. The existing literature on DVFS policies for on-chip networks focuses on using static network parameters like average queue utilization, average return time to memory requests, etc. to decide the new voltage-frequency (V-F) states of the routers. Typically, a DVFS controller would perform the following tasks: namely monitor a suitable network parameter, compute state feedback values based on previous states and target value and update V-F state. Some recent papers on DVFS for on-chip networks are discussed later in the bibliography of this chapter.

Power-Efficient Designs. The second class of technique tries to reduce power consumption by reducing capacitance or switching activity.

The dynamic power of on-chip networks can be reduced by reducing the effective capacitance being switched. Wires dominate network power since wire capacitance is much larger than gate capacitance. Energy-efficient signaling in the form of low-swing [287] and equalized links [314] has been studied in this regard. Router power can also be reduced by reducing the number of pipeline stages, and optimizing the buffers, crossbar, and arbiter circuits/microarchitecture. For instance, SRAMs are more energy-efficient than flip flops and register files for implementing buffers, while matrix-style crossbars are often more efficient than mux-based crossbars. Crossbars can be further segmented [351] or designed with low-swing links [287] to reduce power consumption during traversals. Complex arbiters can be split into multiple simpler arbiters [189, 291] to reduce power consumption further.

Lowering the switching activity is another technique to reduce dynamic power. Clock-gating is a popular method to reduce the amount of switching activity of latches between inactive circuits. For instance, the dynamic power at low-loads in Figure 6.18a is primarily due to the clock, and not actual traffic, providing an opportunity to reduce power. Efficient encoding of the bits being sent from one router to the other could also be exploited to reduce the number of bit-toggles, and thereby dynamic power.

6.6.2   LEAKAGE POWER

At sub-nm technologies, transistors are not ideal switches anymore and leak current even when they are "off." This leads to high power consumption even during periods of low or no activity. Leakage power in on-chip networks has been shown to contribute significantly toward total power consumption at modern technologies, as Figure 6.18a demonstrates. The reason is the large number of latches/flip-flops/SRAMs used for implementing buffers, input VC state, and output credit state.

Power-Gating. Leakage power can be mitigated by power gating. It is a standard technique used across chips today. In this book, we will not go into the circuit details and implications

Power-Gating

of adding power-gating transistors to create power-domains. Instead, we will list some of the challenges that on-chip network power gating solutions need to worry about.

- *What should be the granularity of power-domains?* Candidates for power-domains in an on-chip network could be the various modules in a router (input ports, arbiters, crossbar), or each router by itself, or the entire on-chip network. Fine-grained power gating would be most effective, in principle, but adding power-gating circuitry to hundreds of modules and controlling them is not practical. Most commercial chips today view the entire on-chip network as one power-domain.

- *How to decide which routers to turn on/off?* If the tiles connected to routers are active, the routers will have to be woken up very frequently, adding a lot of latency overhead. Moreover, turning off certain routers may lead to certain key IP blocks, such as the memory controller, becoming inaccessible which is not allowed and something the power management controller needs to take care of.

- *How to handle deadlocks on irregular topologies?* Turning off certain routers make the underlying topology irregular; this can lead to routing deadlocks since certain paths may become inaccessible forcing flits to use other paths that cause cyclic dependencies.

Some recent papers on power gating for on-chip networks are discussed later in the bibliography of this chapter.

6.7   PHYSICAL IMPLEMENTATION

6.7.1   ROUTER FLOORPLANNING

A key step in the backend design flow of a router is floorplanning: determining the placement of the different input and output ports of a router, along with the global allocator modules and the crossbar switch. Figure 6.19 shows two alternative router floorplans for a fairly similar microarchitecture: A 5-port router with virtual channel flow control.

Typically, the allocators (VA) or the crossbar switch traversal (ST) dictate the critical path. Hence, both floorplans optimized their layouts in order to target these two components, but in different ways. Both floorplans use the semi-global metal layers (typically M5, M6 in recent processes) for router-to-router interconnects, but Figure 6.19a drops to the local metal layers for intra-router wiring, such as the crossbar switch, with the inter-router link datapath continuing on the upper metal. Figure 6.19b, on the other hand, has the crossbar switch continuing on the semi-global metal layers.

This leads to a key difference apparent from the floorplan in the placement of the input ports. The placement in Figure 6.19a is fairly intuitive: the north input port is placed close to the north edge of the router, the east input port along the east edge and so on. The placement in Figure 6.19b puts all input ports side by side, on the left side of the switch, to free up the M5 and M6 layers for the crossbar wiring, while ensuring fast access to the allocators.

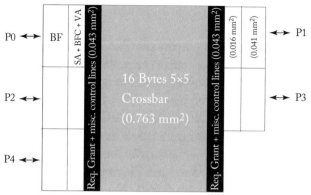

(a) Router Layout from Kumar et al. [208]. BF: Buffer, BFC: Buffer Control, VA: VC Allocator, SA: Switch Allocator. P0: North port; P1: East; P2: West; P3: South; P4: Injection/Ejection.

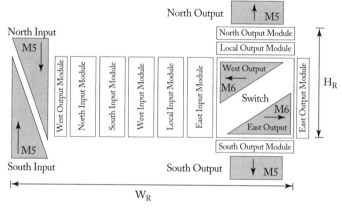

(b) Router Layout from Balfour and Dally [38]. M5 and M6 indicate the metal layers used.

Figure 6.19: Two router floorplans.

To target allocator delay, Figure 6.19a replicates the allocators at every input port, so allocator grant signals will not incur a large RC delay before triggering buffer reads, and the crossbar can also be setup more quickly as control signals now traverse a shorter distance. This comes at the cost of increased area and power. Allocator request signals still have to traverse through the entire crossbar height and width, but their delay is mitigated as that router uses a pipeline optimization technique, advanced bundles, to trigger allocations in advance. Figure 6.19b, however, leverages their use of the semi-global metal layers for the crossbar to place the allocators in the middle, in the active area underneath the crossbar wiring, to lower wire delay to the allocators without replication.

Here, we just aim to illustrate the many possible back-end design decisions that can be made at floorplanning time to further optimize the router design. Note that as routers are just a component of a many-core chip, its floorplan also needs to be done relative to the positions of the network interfaces (NICs), cores and caches.

We will revisit floorplanning in NoC prototypes in Chapter 8. Most recent chip prototypes with NoCs [39, 72, 101] synthesize the entire router as one module rather than hierarchically, letting the CAD tools automatically place the various components of the router within the specified area. For instance, Figure 6.18b plots the area distribution for a state-of-the-art mesh router with 4 VCs at 32 nm [64], designed by letting the CAD tools perform the place-and-route. The buffers and crossbar contribute to over 70% and 20% of the area. In other designs, the router's buffers and arbiters are synthesized and then laid out as one module, leaving aside area for the crossbar which is custom-designed and integrated during final place-and-route [287].

6.7.2   BUFFER IMPLEMENTATION

Router buffer cells can be implemented using flip-flops or generated memory cells (SRAM or register file), depending on the buffer size and access timing requirements. For very small buffers, flip flops suffice and can be readily synthesized without requiring memory generators. Flip flops, however, have much poorer area, power and delay characteristics compared to SRAMs and register files. Between SRAMs and register files, at smaller buffer sizes, register file cells tend to occupy a smaller footprint as they do not require differential sense amplifiers, and support faster access times. However, at larger buffer sizes, SRAMs prevail. The crossover point depends heavily on the specific process and memory cells used, so designers ought to carefully evaluate the alternatives.

6.8   BRIEF STATE-OF-THE-ART SURVEY

Research in efficient NoC design—for performance and/or energy—naturally leads to modifications to the microarchitecture. In this section we discuss recent research that complements the basic microarchitecture design features discussed in this chapter. Chapter 8 presents case studies of NoC prototypes and discusses their choice of router microarchitecture.

Low-latency Routers. Many microarchitectural proposals try to reduce the number of router pipeline stages in order to improve on-chip network latency. Ideas involving circuit-switching [11, 113, 273, 372] and static resource allocation [135, 211] do not require routing and arbitration at every router, reducing router delay. The static resource allocation techniques however add overhead in terms of tables to store the buffer and link bandwidth allocations for every cycle, and only work when the traffic is known in advance, such as in MP-SoCs. Flit reservation flow control [291] tries to recreate this benefit dynamically and reduces buffer turnaround time by letting control flits go ahead on faster links to reserve buffers and free them faster. Static VC allocation [318] removes VC allocation from every router by performing it statically at the source, at the cost of inefficiency in overall VC utilization. Simpler pipelines [139, 185, 255] can be used to reduce the router delay of dynamic networks. Speculative buffer bypassing [101, 208, 209, 210, 242, 285, 287] can be used to reduce delay further by bypassing the buffering stage. All of these techniques also help reduce dynamic and leakage power as there are fewer pipeline registers in each router.

Efficient Management of Virtual Channels. Virtual Channels are precious resources inside each router. Implementing VCs as private vs. shared or shallow vs. deep, come with their pros and cons as Section 6.2.1 discussed. DAMQ [334] and ViChaR [260] allow N buffer slots to act as one deep VC or multiple shallow VCs, depending on the traffic. The challenge with both these designs is that the control overhead of N VCs still needs to be paid, which still adds the critical path, area and power overheads. Centralized buffers [146, 147] instead of per input-port have also been studied to efficiently manage storage within each router, at the cost of adding an additional crossbar at the input stage to drive flits at each input port to the right buffer slots.

Novel Flow Control. There has been research on new flow control protocols for NoCs that demand major modifications to the microarchitecture. Some examples include the use of a ring as the switching fabric within a router rather than a crossbar [3], adding support for multicasts [114, 206], cache coherence support within network routers [16, 111], ability to setup bypass paths across multiple hops [204], and so on. Optimizations to allocation have been proposed including simplifying allocation [149, 367], improving fairness [220] and improving allocation quality [62, 244]. In application-specific MPSoCs, router microarchitectures can also be customized; examples are custom buffer allocation [164], time-slot tables inside routers [135, 211], and asynchronous routers [41, 178].

Novel Circuits. A few papers have proposed novel circuits for microarchitectural components. Low-swing links inside crossbars [287] have been proposed to reduce power consumption. Arbitration support within the crossbar has also been proposed [315]. Both these crossbars have been prototyped with test chips and will be discussed in Chapter 8. Novel micro-architectures and circuits for buffers have also been proposed in the form of elastic buffers [243] to improve bandwidth and reduce power consumption.

Low Power Routers. Wang et al. [351] present circuit and micro-architectural optimizations such as a write-through buffer and a segmented crossbar to reduce router power. RoCo [189] decouples traffic going along X and along Y, effectively splitting the arbiters and crossbar into smaller and simpler units, that are more power efficient. Bufferless routing [148, 253] makes a case for bufferless on-chip networks with either mis-routing in case of contention or packet dropping. These help reduce on-chip power consumption as long as the number of mis-routes or dropped packets, which increases power consumption of links, is not too high. Considering novel buffer organizations such as centralized buffers can also reduce router power consumption [146]. Simplified routers have been designed to enable lower power and area consumption [185].

DFVS in NoCs. Research on DVFS in NoCs has explored various heuristics for V-F assignment. One set of works use NoC metrics to tune voltage and frequency, such as target throughput [58, 71], buffer utilization [51, 250], energy consumption [269], and errors [24]. Another set uses runtime performance of application workloads for V-F assignment. This is done by observing system-level metrics such as coherence messages [153], L1/L2 misses [52, 369], and memory-access density [370]. Online learning is also being employed to predict V-F settings [361].

Power Gating of NoC Routers. Power gating of NoC routers has been studied recently [67, 69, 70, 100, 283, 306]. While turning off the router provides benefits in terms of leakage energy reduction, it can lead to routing deadlocks due to absence of certain paths and requires additional support in the form of deadlock-free Up/Down routing within the NoC [67, 306].

CHAPTER 7

Modeling and Evaluation

Modeling all aspects of the on-chip network architecture (topology, routing, flow-control, and router-microarchitecture) either in simulation, or in real-design, is crucial for the design-space exploration and validation of on-chip networks. The resulting on-chip network is evaluated for performance (with synthetic and/or real traffic), power, and area.

7.1 EVALUATION METRICS

On-chip networks are typically characterized/evaluated by their performance (latency and throughput), energy consumption, and area footprint.

7.1.1 ANALYTICAL MODEL

Latency. The latency of every packet in an on-chip network can be described by the following equation:

$$T_{Network} = T_{wire} + T_{router} + T_{contention}$$
$$= H \cdot t_{wire} + (H + 1) \cdot t_{router} + \sum_{h=1}^{H+1} t_{contention}(h),$$

where H is the average hop count through the topology, t_{router} is the pipeline delay through a single router, t_{wire} is the wire delay between two routers, and $t_{contention}(h)$ is the delay due to contention between multiple messages competing for the same network resources at a router h-hops from the start. A factor of $H + 1$ is considered for router power and contention since a packet traverses the input router prior to the first hop through the network. t_{router} accounts for the time each packet spends in various stages at each router as the router coordinates between multiple packets; depending on the implementation, this can consist of one to several pipeline stages as discussed in Chapter 6. t_{router} and t_{wire} are design-time metrics. They can be used to determine a lower-bound on the latency of any packet. H and $t_{contention}(h)$ are runtime metrics that depend on traffic.

Throughput. The bisection bandwidth, defined earlier in Chapter 3, is a design-time metric for the throughput of any network. As a reminder, it is the inverse of the maximum load across the bisection channels of any topology. Ideal throughput assumes perfect flow control and perfect load balancing from the routing algorithm. The actual throughput at saturation, however, might vary heavily, depending on how routing and flow control interact with runtime traffic.

Throughput higher than the bisection bandwidth can be achieved if traffic does not go from one end of the network to the other over the bisection links. However, often times, the achieved saturation throughput is lower than the bisection bandwidth. A deterministic routing algorithm, such as XY, might be unable to balance traffic across all available links in the topology in response to network load. Heavily used paths will saturate quickly, reducing the rate of accepted traffic. On the other hand, an adaptive routing algorithm using local congestion metrics could lead to more congestion in downstream links. The inability of the arbitration schemes inside the router to make perfect matching between requests and available resources can also degrade throughput. Likewise, limited number of buffers and buffer turnaround latency can drive down the throughput of the network.

Energy. The energy consumed by each flit during its network traversal is given by:

$$
\begin{aligned}
E_{Network} &= H \cdot E_{wire} + (H+1) \cdot E_{router} \\
&= H \cdot E_{wire} + (H+1) \cdot (E_{ST} + E_{BW} + E_{BR} + E_{RC}) + \\
&\quad \sum_{h=1}^{H+1} t_{contention}(h) \cdot (E_{VA} + E_{SA}),
\end{aligned}
$$

where E_{BW}, E_{RC}, E_{VA}, E_{SA}, E_{BR}, and E_{ST} is the energy consumption for buffer write, route computation, VC arbitration, switch arbitration, buffer read, and switch traversal, respectively. E_{RC} and E_{VA} are only consumed by the head flit. The relative contribution of these parameters is topology and flow control specific. For instance, a high-radix router might have a larger E_{ST} and E_{wire}, but lower H. Similarly, a wormhole router will not consume E_{VA}. Contention at every router determines the number of times a flit may need to perform VA and SA before winning both and getting access to the switch. E_{VA} and E_{SA} depends on the specific allocator implementation.

Area. The area footprint of an on-chip network depends on the area of routers.

$$
\begin{aligned}
A_{Network} &= N \cdot (A_{router}) \\
&= N \cdot (p \cdot v \cdot A_{VC} + p \cdot A_{RouteUnit} + p \cdot A_{Arbiter\_inport} + p \cdot A_{Arbiter\_outport} \\
&\quad + A_{Crossbar}),
\end{aligned}
$$

where N is the number of routers (assuming all of them are homogeneous input buffered designs), p is the number of ports, and v is the number of VCs per input port. A_{VC} is the area consumed by the buffers and control for each VC, which in turn depends on its implementation, as Chapter 6 discussed. This equation assumes a separable switch allocator design; $A_{Arbiter\_inport}$ represents the area of all the arbiters at each input port, and $A_{Arbiter\_outport}$ represents the area of all the arbiters at each output port.

Wires do not directly contribute to the area footprint as they are often routed on higher metal layers above logic; the link drivers are embedded within the crossbar while the link receiver is within the input VC.

7.1.2   IDEAL INTERCONNECT FABRIC

Ideal latency. The lowest latency (or ideal latency) that can be achieved by the interconnection network is one that is solely dedicated to the wire delay between a source and destination. This latency can be achieved by allowing all data to travel on dedicated pipeline wires that directly connect a source to a destination. Wire latency assumes the optimal insertion of repeaters and flip-flops. The latency of this dedicated wiring would be governed only by the average wire length D between the source and destination (assumed to be the Manhattan distance), packet size L, channel bandwidth b, and propagation velocity v:

$$T_{ideal} = T_{wire} = \frac{D}{v} + \frac{L}{b}.$$

The first term corresponds to the time spent traversing the interconnect, while the second corresponds to the serialization latency for a packet of length L to cross a channel of bandwidth b. Ideally, serialization delay could be avoided with very wide channels. However, such a large number of wires would not be feasible given the projected chip sizes.

Ideal throughput. The ideal throughput depends solely on the bandwidth provided by the topology. It can be computed by calculating the load across all links in the topology for a particular traffic pattern with a specific routing algorithm, and taking the inverse of the maximum load link.

Ideal energy. The energy expended to communicate data between tiles should ideally be just the energy of interconnect wires as given by

$$E_{ideal} = E_{wire} = \frac{L}{b} \cdot D \cdot E_{wire},$$

where D is again the distance between source and destination and E_{wire} is the interconnect transmission energy per unit length.

7.1.3   NETWORK DELAY-THROUGHPUT-ENERGY CURVE

We simulate a state-of-the-art virtual channel 8×8 mesh network using the Garnet [14] network-on-chip simulator. The routers implement some of the pipeline optimizations discussed in Chapter 6, namely lookahead routing, VC selection, and lookahead bypass, leading to a single-cycle router pipeline at every hop (2-cycles per hop) in the best case. The full set of network parameters is given in Table 7.1. We plot the average latency vs. injection rate for two synthetic traffic patterns: uniform random traffic and bit-complement traffic in Figure 7.1. These patterns are described later in Table 7.2. We send 1-flit packets at increasing injection rates from each node. The saturation throughput is defined as the injection rate where latency becomes 3× that at low loads. We also plot the ideal latency and throughput lines on these graphs. For the ideal latency, we assume a dedicated repeated wire between each pair of tiles with a propagation velocity v = *tile_size/cycle*, i.e., 1 cycle to cross each tile (i.e., hop). The average value for ideal latency is simply the average hop count for each traffic, plus an additional

Table 7.1: State of the art network simulation parameters

Parameter	Value
Technology	45 nm
V_{dd}	1.0 V
Frequency	2 GHz
Topoloty	8-ary 2-mesh
Routing	Dimension-ordered (DOR)
Traffic	Uniform Random and Bit Complement
Router pipeline depth	1
Number of router ports	5
VCs per port	4
Buffers per port	4 (1-per VC)
Flit size (channel width)	128 bits
Link length	1 mm

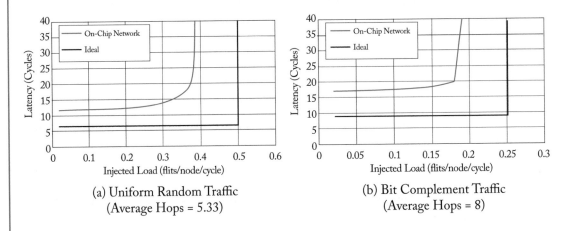

(a) Uniform Random Traffic
(Average Hops = 5.33)

(b) Bit Complement Traffic
(Average Hops = 8)

Figure 7.1: Latency vs. injected traffic for a 8 × 8 mesh on-chip network.

hop to get from the final router to the destination NIC. The ideal throughput is computed using maximum channel load at the bisection links.

At low-loads, the state-of-the-art design is close to the ideal latency, the gap is due to the additional 1-cycle router delay at every hop in the former. This small gap is due to the pipeline optimizations incorporated into the design. A 5-stage pipeline at every router would increase this gap significantly, leading to system-level performance penalties. At very high loads, the gap increases due to contention. The state-of-the-art VC router delivers about 80% throughput of the

Table 7.2: Synthetic traffic patterns for $k \times k$ mesh

Source (binary coordinates): $(y_{k-1}, y_{k-2}, ..., y_1, y_0, x_{k-1}, x_{k-2}, ..., x_1, x_0)$			
Traffic Pattern	Destination (binary coordinates)	Avg Hops (for $k + 8$)	Throughput (for $k = 8$) (flits/nodes/cycle)
Bit-Complement	$(\overline{y}_{k-1}, \overline{y}_{k-2}, ..., \overline{y}_1, \overline{y}_0, \overline{x}_{k-1}, \overline{x}_{k-2}, ..., \overline{x}_1, \overline{x}_0)$	8	0.25
Bit-Reverse	$(x_0, x_1, ..., x_{k-2}, x_{k-1}, y_0, y_1, ..., y_{k-2}, y_{k-1})$	5.25	0.14
Shuffle	$(y_{k-2}, y_{k-3}, ..., y_0, x_{k-1}, x_{k-2}, x_{k-3}, ..., x_0, y_{k-1})$	4	0.25
Tornado	$(y_{k-1}, y_{k-2}, ..., y_1, y_0, x_{k-1 + [\frac{k}{2}] - 1}, ..., x_{[\frac{k}{2}] - 1})$	3.75	0.33
Transpose	$(x_{k-1}, x_{k-2}, ..., x_1, x_0, y_{k-1}, y_{k-2}, ..., y_1, y_0)$	5.25	0.14
Uniform Random	$random(\ )$	5.25	0.5

ideal for both traffic patterns. The 20% gap is due to inefficiencies in routing and arbitration that lead to a loss in link utilization at high loads. Simpler router designs will increase this throughput gap significantly; wormhole flow control without virtual channels will saturate much earlier than the curve shown. A small number of buffers will also reduce the saturation throughput.

Figure 7.2 plots the energy consumption of an ideal network and a state-of-the-art baseline network, using the DSENT [328] energy models. This baseline architecture incorporates many energy-efficient microarchitectural features but still significantly exceeds the energy consumed solely by wires. This gap exists due to the additional buffering, switching, and arbitration that occurs at each router; the gap widens until the network saturates.

7.2   ON-CHIP NETWORK MODELING INFRASTRUCTURE

There are several on-chip network modeling infrastructures developed by computer architecture and network-on-chip researchers, that can be used to study novel network topologies, routing algorithms, flow-control methods, router microarchitectures, and emerging link technologies. Some of these exist as standalone tools, while others are part of larger full-system simulators that model CMPs or MPSoCs, as described earlier in Chapter 2 to drive real traffic.

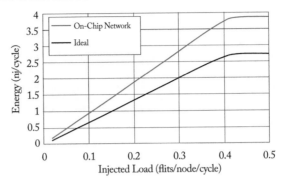

Figure 7.2: Network energy with uniform random traffic in a 8 × 8 mesh on-chip network.

7.2.1   RTL AND SOFTWARE MODELS

Modeling on-chip networks in RTL and using Cadence/Synopsys/MentorGraphics tools for RTL simulation gives the most accurate network implementation, and the most cycle-accurate timing information. The on-chip network RTL serves as both a model as well as the final design and is a common design space exploration methodology in industry.

The increased on-chip logic and memory capacities of modern FPGAs allow the entire on-chip system to be implemented on a single device. Compared to software, FPGA-based on-chip network emulators can reduce simulation time by several orders of magnitude [129, 201, 359]. These dramatic speedups are possible because the emulator is constructed by laying out the entire on-chip network on the FPGA, allowing the hardware to exploit all available fine and coarse grain parallelism between the emulated events in the on-chip network.

Once the end points of the on-chip network, namely the multicore system is also included, simulating this entire system in RTL becomes highly intractable. As a result, software (e.g., C++) simulators are used extensively for design space exploration for network design, as well as co-design of the network with the rest of the memory sub-system.

The topology, routing algorithms, flow-control, and router microarchitecture can be modeled in various degrees of detail, trading-off accuracy for simulation time. It is not recommended to fudge the network model however, especially in large multicore simulations, since on-chip network latency and bandwidth might directly affect the performance of the full distributed CMP/MPSoC, and can only be captured by modeling cycle-by-cycle contention within the network.

Section 7.6 summarizes some of the state-of-the-art on-chip network simulators.

7.2.2   POWER AND AREA MODELS

Power consumption is a first-order design constraint in systems today. Early-stage estimation of on-chip network power is becoming crucial in order to budget power across the various subsystems. The key contributors to on-chip network power are the buffers, crossbar switches, and the tile-to-tile links, as discussed earlier in Chapter 6.

RTL models of routers [118, 281] can be synthesized, placed, and routed for accurate area estimates. The resulting netlists can be back-annotated with activities from a RTL simulation, to give accurate power estimates. However, this rigorous approach often becomes complex and even infeasible when modeling a full CMP, with cores, caches, and the on-chip network. The reasons include unavailability of RTL for every component, access to standard cells for advanced nodes and simulation speed. Thus, researchers have developed detailed software models for the energy consumption and area footprint of various network components to enable designers to size or optimize these at design-time to fit within the power and area budgets for the expected input activities.

DSENT [328] is a timing-driven on-chip network area and power modeling tool for both electrical and optical on-chip networks. It provides a technology-portable set of standard cells from which larger electrical components such as router buffers, arbiters, crossbars, and links are constructed. Given a foundry's process design kit (PDK) and the design's frequency constraint, DSENT applies (1) timing optimization to size gates for energy-optimality and (2) expected transition propagation to accurately gauge the power consumption. DSENT is available for download as a standalone tool [251] and is also released as part of gem5 [49]. It replaces the older Orion 2.0 [174] model which modeled on-chip network components characterized against post-layout implementations at 65 nm. McPAT [224] is an integrated power, area, and timing modeling tool for multicores that models the power of the processors, caches and the interconnect. It is validated against published chips in 180 nm, 90 nm, and 65 nm, and uses DSENT for the on-chip network power models.

7.3   TRAFFIC

The traffic through the on-chip network depends on the kind of system it has been plugged into and the overlaying communication protocol. Some of the common communication protocols were described in detail in Chapter 2. Here we discuss how the communication protocol prescribes the modeling and evaluation of on-chip networks. For the purpose of illustration, we consider shared memory systems where the on-chip network interconnects the memory subsystem (L1, L2, directory, memory controllers, etc.) and transfers cache-coherence traffic.

7.3.1   MESSAGE CLASSES, VIRTUAL NETWORKS, MESSAGE SIZES, AND ORDERING

Message Classes. The overlaying coherence protocol defines various message classes. For instance, in most protocols there are at least two message classes: *request* and *response*. Directory-based cache coherence protocols often use four message classes: *request*, *forward*, *response*, and *unblock*.

Virtual Networks. A potential deadlock can occur in the protocol if a request for a line from a L2 cache is unable to enter the network because the L2 is waiting for a response for a previous request, while the response is unable to reach the L2 since all queues in the network are full of such waiting requests. To avoid such deadlocks, protocols require messages from different message classes to use different set of queues/buffers within the network. This is implemented by using **virtual networks (vnets)** within the physical network. Virtual networks are identical to VCs in terms of their implementation: all vnets have separate buffers but multiplex over the same physical links. In fact many works on coherence protocols use the term virtual channels to refer to virtual networks. However, in this book we will strictly adhere to using the term virtual networks or vnets to refer to protocol-level message classes. The number of vnets is thus fixed by the protocol. Each vnet, on the other hand, can have one or more VCs within each router, to avoid head-of-line blocking or routing deadlocks, as discussed earlier in Chapter 5.

Message Sizes. The size of each message depends upon the protocol and the message type. Control messages, such as requests/forwards/unblocks, are often shorter as they need to carry the address and some header information, while data messages, such as responses, are cache-line sized with some extra header information. On-chip networks often try to set the channel sizes such that control messages fit within one flit, while data messages might fit in one or more flits. For example, suppose a cache-coherent system uses 64 b addressing, 64 B cache lines, and 16 b headers; if the on-chip network's channel width is 128 bits, then control packets would fit within 1-flit, while data packets would fit in 5-flits.

Point-to-Point Ordering. Certain message classes (and thus their vnets) require point-to-point ordering in the network for functional correctness. This means that two messages injected from the same source, for the same destination, should be delivered in the order of their issue. On-chip networks can implement point-to-point ordering for flits within ordered vnets by (i) using deterministic routing and (ii) using FIFO/queuing arbiters for switch arbitration at the input port at each router. The first condition guarantees that two messages from the same source do not use alternate paths to the same destination as that could result in the older message getting delivered after the newer one if the former's path has more congestion. The second condition guarantees that flits at a router's input port leave in the order in which they came in.

7.3.2   APPLICATION TRAFFIC

In MPSoCs, an application's communication task graph, such as the one shown earlier in Figure 3.8a, determines the traffic flows between various IPs connected via an on-chip network. Traffic models can be extracted based on average traffic flowing between cores [156, 161]. This helps drive customized network topologies and mapping algorithms [256] for traffic from the class of applications the MPSoC runs. The edges on the task graph determine the throughput requirement from network links while the physical number of routers between communicating IPs on chip and the contention between flows mapped over the same links determines the network latency.

In multicore systems, which are typically shared memory, the sharing behavior of the application and the kind of coherence protocol used (snoopy vs. limited-directory vs. full-bit directory) determines the traffic through the network. For an N-core chip, the communication patterns of protocols can be classified into 1-to-1, 1-to-M, and M-to-1 where M refers to multiple sources or destinations ($1 < M <= N$). 1-to-1 or *unicast* communication occurs in unicast requests/responses exchanged between cores. 1-to-M or *multicast* communication occurs in broadcasts and multicast requests. M-to-1 or *reduction* communication occurs in acknowledgements or token collection in protocols to maintain ordering. Full-bit directories track the state of every sharer and use unicasts and precise multicasts to maintain coherence. Snoopy protocols and limited-directory (i.e., directory protocols with limited tracking of sharers) protocols trade-off directory storage for higher network traffic in the form of broadcasts, multicasts and reductions.

Network traffic from an application also depends on cache sizes and cache hierarchies, and how much of the application's working set fits. L1 and L2 sizes determine their miss rates which in turn determines injection rates into the network. Distributed shared L2s also have higher network traffic compared to private L2s per core since every L1 miss has to traverse the on-chip network to reach the home node in the former. Memory bound traffic also traverses the on-chip network to reach memory controllers, and can lead to Quality of Service issues if certain cores are always able to get faster access to the memory controller.

Faithfully modeling these different aspects of on-chip network traffic is important for properly evaluating on-chip network performance and power consumption. On-chip network traffic can be injected either by using traces, or by running full-system simulations.

Trace-driven Simulation. Researchers often use traces of network injections from applications running on a real system or a full-system simulator. Network traces provide a fairly realistic way of exploring the effectiveness of proposed on-chip network designs, but clearly, it should be noted that their characteristics depend heavily on the simulated many-core platform. The number of cores/IP blocks, the memory hierarchy, the number of memory controllers, etc. significantly influence the network trace. The lack of feedback effects when using network traces also impacts the accuracy. For instance, faster on-chip networks than the ones on which traces were collected could lead to pathological scenarios such as responses getting delivered before their requests were injected. Tracking or inferring dependencies between packet is important to being able to replay the trace correctly [263]. Netrace [155] is a set of tools and traces designed to enhance the performance and fidelity of traditional trace-based on-chip network simulations by adding dependency tracking within the trace-based simulation framework.

Full-system Simulation. Full-system evaluations provide the most accurate traffic movement within the network, as they model the entire system (cores, caches, coherence protocol, network, and memory) in detail and boot an OS on which the application is run. However, these simulations take up substantial simulation time. Benchmark suites exist for shared memory and MPI applications, which in turn stress the on-chip network to varying degrees depending on the application thread mapping across cores, sharing pattern, coherence protocol (if shared memory), cache sizes, core model (in-order or out-of-order), and so on during the full-system simulation. Examples include SPLASH-2 [362], PARSEC [48], and Rodinia [63].

7.3.3 SYNTHETIC TRAFFIC

Synthetic traffic patterns help in characterizing and debugging on-chip networks. In most standard synthetic patterns, all sources inject with a uniform random injection rate (without bursts), while the destination coordinates depend on the traffic pattern. Table 7.2 lists some common synthetic traffic patterns used for studying a mesh network, along with their average hop-counts and theoretical throughput with XY routing. The theoretical throughput or capacity is the injection rate at which some link(s) in the mesh is (are) sending 1 flit every cycle.[1] These values are computed using the maximum channel load procedure described earlier in Chapter 3. For each traffic pattern, the load across all links is calculated assuming XY routing. The link that is the most heavily loaded is determined. The theoretical throughput is the inverse of this load. This is the best a topology can do, with perfect routing, flow control and microarchitecture.

Synthetic traffic patterns are useful for characterizing the latency and throughput of the on-chip network, by plotting the latency vs. throughput graphs for the proposed on-chip network and comparing them with those of the baseline and an ideal. Design-space explorations can be done with this data to see if the desired latency and throughput can be met with fewer buffers, VCs, flit sizes, channel widths and so on.

Beyond traditional synthetic traffic patterns, there are also synthetic traffic generators that mimic application and cache coherence traffic. SynFull [34] focuses on synthetically reproducing the traffic dependences in cache coherence traffic and the fluctuations in traffic volume across different phases of an application for general-purpose CMP workloads. APU-SynFull [371] extends this work to focus on heterogeneous CPU-GPU architectures with more complex coherence patterns and more bursty traffic.

In general, a combination of synthetic traffic flows, which exercise and test the limits of a proposed approach; real network traces, which give an idea of the effectiveness of an approach; and full-system simulations, which more accurately evaluate the approach in a specific system; should be used to have a good understanding of the pros and cons of a proposed technique.

7.4 DEBUG METHODOLOGY

There is a very tight set of dependencies between the distributed resources (buffers, VCs, and links) of an on-chip network. Even if only one VC or buffer gets indefinitely blocked due to an incorrect flow-control handshake, it can easily bring the entire on-chip network to a standstill in a matter of cycles due to a cascading effect on other dependent VCs, which in turn would crash the entire system. Most bugs manifest as an on-chip network deadlock and it becomes difficult

---

[1]Table 7.2 shows that uniform random traffic offers the highest throughput, since it saturates when the bisection links of the mesh are fully occupied. For traffic patterns that saturate other links, throughput is lower.

to decipher where the problem is. We list a set of tips for novice researchers for debugging on-chip network optimizations.

- Once you add your desired topology/routing/flow-control/microarchitecture optimization, run your simulator in a standalone mode and make sure *one* flit is successfully delivered from the a fixed source to a fixed destination. Most bugs can be found and fixed here by probing the intermediate routers and states.

- Repeat the same experiment for multiple flits from fixed/variable sources and destinations. This can be done by running one of the standard synthetic traffic patterns, system traces, or a user-specified pattern.

- Once the on-chip network is robust across the synthetic traffic patterns, it can be plugged into a full-system with real traffic. Common errors at this stage include incorrect ordering or virtual network violations within the on-chip network, which leads to flits getting delivered without the on-chip network deadlocking, but may lead to coherence protocol violations.

- To characterize the effectiveness of the technique, it is often useful to also plot the ideal behavior of the on-chip network as well, which can be modeled by implementing an impractical but ideal fully connected network giving the lowest delay and highest throughput without contention.

7.5   NOC GENERATORS

NoC RTL generators are provided by commercial vendors and academic researchers for plug-and-play into CMPs and/or MPSoCs. These generators use a library of modularized components to build routers with varying number of ports, data widths, and buffer depths. Some of these provide application-specific synthesis for heterogeneous SoCs, while some generate homogeneous NoCs for multicores with different topology and routing algorithms. Detailed information on application-specific NoC synthesis can be found in Benini and De Micheli [45].

Commercial. ARM's CoreLink [28] interconnect generates buses and mesh networks tailored to ARM Cortex and Mali cores in both cache-coherent CMPs and mobile SoCs. FlexNoC [29] by Arteris is a proprietary NoC generation tool to connect IPs implementing any combination of AMBA®ACE™, ACE-Lite, AXI™, AHB™, AHB-Lite, APB™, OCP, and PIF protocols. It lets the designer specify a task graph which translates into a network topology. The designer can then add "links" to share multiple flows. A library of RTL components is used to create the actual NoC. Functional and timing simulations can be performed on the designs. The tool flow also performs automated pipelining to meet timing. SonicsGN [323] is similarly a configurable NoC generator from Sonics for heterogeneous SoCs. It scales from high-throughput networking chips operating in GHz to low-power and low-latency IoT wearables operating at MHz frequencies.

Academic. ×pipes [169, 327] is a NoC generator in SystemC, with a library of configurable NoC components. It performs topology synthesis [256] based on the target application. DRNoC [202] and Connect [281] are NoC generators optimized for FPGAs. DRNoC generates application-specific NoCs taking the communication task graph as an input. Connect uses Bluespec System Verilog (BSV) [261] for the implementation and generates synthesizable Verilog of user-specified topologies and design parameters. It supports a web-based graphical user interface so that users can obtain network designs with various topologies easily. The generated verilog is available for users, but not the BSV source code due to licensing issues. Open SoC Fabric [118] provides an NoC generator written in Chisel [33]. This generator supports 2-D mesh and flattened butterfly networks of arbitrary sizes. The source code is open, and users can freely edit the source code and re-compile it because Chisel is an open-source language. The NoC System Generator [59] receives a specification of NoCs in an XML file format and produces VHDL for this specification. It supports 2-D and 3-D mesh topologies. OpenPiton [39] is an open-source many-core system generator, and it has used for fabricated ASIC chips and an FPGA implementation that runs full-stack Linux. OpenPiton contains a NoC structure to support the cache coherence, memory, and inter-core interrupt traffic of the SPARC cores it employs. OpenSMART [213] is a recent open-source NoC generator that provides both BSV and Chisel implementations for mesh and SMART [204] routers.

7.6 BRIEF STATE-OF-THE-ART SURVEY

Many tools are available to designers and researchers today to model, evaluate, and refine on-chip network designs. In the MPSoC domain, Æthereal [134] and Nostrum [211] provide high level models that allow for iterative design refinement, while cycle-accurate simulators such as MPARM [42] can be used to do design space exploration. In the CMP domain, many open software simulation frameworks exist for modeling on-chip networks, either in a standalone manner, or as part of a full-system. There also exist on-chip network modeling frameworks in RTL—either Verilog, or higher level languages like Bluespec System Verilog [261] and Chisel [33]. Table 7.3 lists some of the state-of-the-art frameworks that provide cycle-accurate on-chip network simulation. An evolving list of simulators is also maintained by Cristinel Ababei (Marquette University), Partha Pande (Washington State University), and Sudeep Pasricha (Colorado State University) [272]. Some full-system simulators such as SESC [303], zsim [307] and others do not model on-chip networks in a cycle-accurate manner for simulation speed, and instead use fixed or probabilistic delays. Apart from simulators, analytical models for on-chip network performance analysis across various topologies and microarchitectures have also been designed [266, 292].

Table 7.3: State-of-the-art on-chip network simulators

Simulator	Language	Environment
Garnet [14, 203]	C++	Standalone + Full-System (gem5 [49])
Booksim2 [170]	C++	Standalone + Full-System (gpgpusim [36])
Topaz [1]	C++	Standalone
Flexsim1.2 [144]	C++	Standalone
SuperSim [277]	C++	Standalone
NOCulator [264]	C++	Standalone
OpenSMART [213]	BSV, Chisel	Standalone
OpenSoC [118]	Chisel	Standalone
CONNECT [281]	BSV	Standalone (optimized for FPGA)
NoCem [309]	VHDL	Standalone (optimized for FPGA)
Agate [66]	C++	Network-only (with Garnet) + Full-System (gem5 [49])
NoCGEN [59]	VHDL	Standalone
DART [349, 350]	Verilog	Standalone

CHAPTER 8

Case Studies

Over the past decade, on-chip networks have been driving real multicore chips across commercial products and research prototypes. We discuss a few of them here as case studies, focusing on the system they are interconnecting and the design specifications. We highlight the topology, routing algorithm, flow control and router microarchitecture and relate these designs back to fundamental concepts presented in earlier chapters; *however, in some cases, limited public information is available about these chips so their treatment may not be complete.* Table 8.1 summarizes the features of all the chips discussed in this chapter. Case studies are presented in chronological order starting with the most recent.

8.1 MIT EYERISS (2016)

The Eyeriss ASIC from MIT [72] is an accelerator for deep convolutional neural networks, with 168 processing elements connected via multiple NoCs. Convolutional neural networks (CNNs) have demonstrated unprecedented levels of accuracy on vision-based machine learning tasks such as object recognition and detection. Eyeriss is part of an emerging trend of accelerator IPs being developed to provide orders of magnitude performance and energy benefits compared to general purpose cores. This trend is fueled in part because of (a) the end of performance scaling due to limits of Dennard's scaling, (b) the dark silicon problem where chips have more transistors than a system can fully power at any point in time, and (c) the emergence of new classes of applications such as deep learning that require a lot of parallel big-data processing. For instance, the computational complexity of CNN comes from the high-dimensional convolution operations (i.e., multiply accumulates), which account for over 90% of the operations.

Eyeriss is built using 168 PEs that communicate with one another directly, rather than via memory, and naturally requires an on-chip network. However, as the communication pattern is known apriori for each layer of the CNN, a lightweight NoC with configurable switches is used, instead of the general-purpose ones described so far in this book. The NoC and PEs are configured before the start of each layer.

The diephoto of Eyeriss is shown in Figure 8.1a. Eyeriss has a global SRAM buffer (GB) which multicasts input feature maps (ifmap) and filter weights to a set of PEs. The logical dataflow is shown in Figure 8.1b: filter weights are multicast along the row, ifmaps are multicast along the diagonal. and partial sums (psum) generated by each PE are sent to their immediate neighbor above. Depending on how this logical dataflow is cut or folded to be mapped on the

Table 8.1: Chip prototypes

Chip	Year	Tech Node	Topology	Route	Flow Control	Router Stages
Eyeriss (MIT) [72]	2016	65 nm	12×14 Mesh	Source	None	0
Piton (Princeton) [39]	2015	32 nm	5×5 Mesh	XY	Wormhole	1, 2
Xeon Phi (Intel) [321]	2015 2012	14 nm 22 nm	6×6 Mesh Ring	YX	Rotary †	1, 2
Anton 2 (DESRES) [338]	2014	40 nm	4×4 Mesh + 2 skip links	Y- → X+ → X- → Y+	VCT	4
SCORPIO (MIT) [101]	2014	45 nm	6×6 Mesh	XY	VC	3, 1
Sparc T5 (Oracle) [341]	2013	28 nm	8×9 Xbar	NA	NA	NA
Swizzle Switch (UMichigan) [315]	2012	45 nm	64×64 Xbar	NA	NA	4
Broadcast NoC (MITY) [287]	2012	45 nm	4×4 Mesh	XY Tree	VC	3, 1
3D Maps (Georgia Tech) [184]	2012	130 nm	8×8 Mesh	Compiler	None	0
Multicast NoC (KAIST) [191]	2010	130 nm	Hier. Ring	Source	Wormhole	4
Intel SCC (Intel) [160]	2009	45 nm	6×4 Mesh	XY	VC	3
ASAP (UC Davis) [339]	2009	65 nm	13×13 Mesh	Source	Circuit Switched	NA
TilePro 64 (Tilera)[356]	2008	90 nm	6 8×8 Meshes	XY Source	Wormhole Ckt-Switch	1, 2
STNoC (ST) [82]	2008	65 nm	Spidergon	Across-First	VC	1, 2
Intel Teraflops (Intel) [158]	2007	65 nm	8×10 Mesh	XY	VC	5
Cell (IBM) [195]	2005	65 nm	Ring	Shortest	Rotary†	1

†Rotary Rule: Traffic on the ring has higher priority than injected traffic.

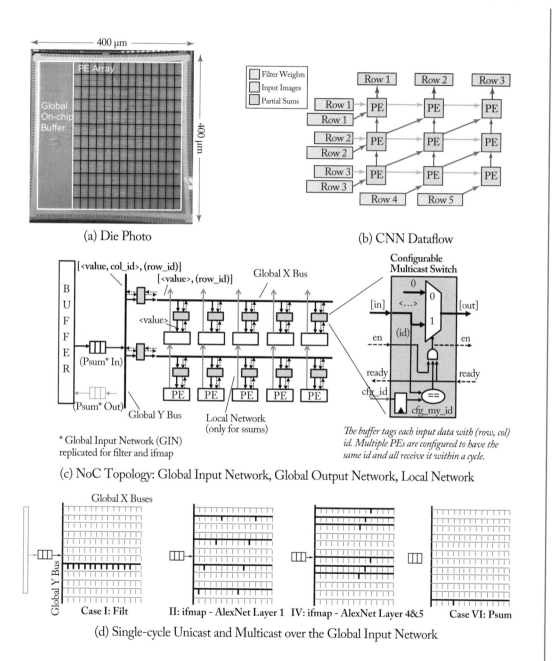

(a) Die Photo

(b) CNN Dataflow

The buffer tags each input data with (row, col) id. Multiple PEs are configured to have the same id and all receive it within a cycle.

* Global Input Network (GIN) replicated for filter and ifmap

(c) NoC Topology: Global Input Network, Global Output Network, Local Network

(d) Single-cycle Unicast and Multicast over the Global Input Network

Figure 8.1: MIT Eyeriss: 12 × 14 mesh [72].

array, and the size of the filters, the multicast pattern may not be as uniform. It is, however, static during the run of each CNN layer.

The Eyeriss NoC comprises three separate networks, shown in Figure 8.1c.

(1) Global Input Network (GIN): The GIN is optimized for a single-cycle multicast from the GB to a group of PEs that receive the same filter, ifmap, or psum. The GIN is built using one Y bus and 12 X buses, one per row. A separate GIN is implemented for each of the three data types (filter, ifmap and psum) in order to provide sufficient bandwidth from the GB to the PE array. The filter and psum GIN have a bus width of 64 b to deliver 4 contiguous words to a PE in a cycle. The ifmap GIN has the data bus width of 16 b.

(2) Global Output Network (GON): The GON has the same architecture as the GIN, except that it is used for reading psums from PEs and sending them to the GB.

(3) Local Network (LN): The LN is a set of point-to-point Y-directional links between the PEs for transfer of psums.

Together these networks logically create a 14×12 mesh topology. The X-links of this mesh are used to deliver data from the global buffer to the PEs, while the Y links are used for local communication between the PEs. The GIN and the GON have a 5-bit tag field, in additional to the data. A PE can get its input psums either from the psum GIN or LN. The selection is static within a layer, which is controlled by the scan chain configuration bits and only depends on the dataflow mapping of the CNN shape.

There is one route between any two communicating entities (GB to PE, PE to GB, and PE to PE) in Eyeriss. Multicasts from the GB use a tree from the root (i.e., Y bus) to the leaf PEs. PEs use the direct local Y link to send psums to their neighbor, or the X bus on the GON to communicate with the GB. This is configured statically.

All PEs that need to receive the same data (filter or ifmap) during multicast are configured with the same ID during the configuration phase. The GB tags each multicast message with the ID of the receiving PEs. All switches/controllers on the GIN multicast the data to those PEs whose IDs match the tag. The multicast takes a single-cycle. This is shown in Figure 8.1d. On-off flow-control is used. A multicast over the GIN is sent out only if all PEs have a buffer available. Delivery to a subset of PEs is not allowed. On the GON, the GB sets the id of the PE it wants to receive from in the tag field. Only this PE is allowed to send the psum. Unlike PEs that receive filters and ifmaps, the PEs that generate output psums have unique ids, so there is never a conflict. Thus no arbitration is required for any of the global buses in the networks since there is only one sender at any point of time.

The switches in the NoC are configurable and bufferless. Their role is to simply pass the data forward if the tag matches. Figure 8.1c shows the microarchitecture.

8.2   PRINCETON PITON (2015)

The OpenPiton Processor from Princeton [39] is a manycore research framework; it was demonstrated with a 25-core tiled processor ASIC prototype called Piton at 32 nm, running at 1 GHz.

The manycore processor uses OpenSPARC T1 cores, and provides a memory subsystem inter-connected via three NoCs. A high level overview is shown in Figure 8.2.

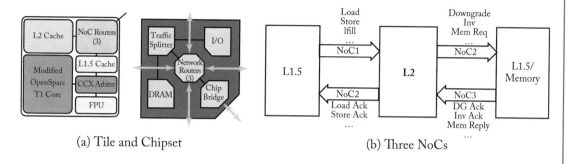

(a) Tile and Chipset   (b) Three NoCs

Figure 8.2: Princeton Piton manycore processor: 6 × 6 mesh [39].

The interface to the OpenSPARC T1's L1 is the OpenSPARC CCX (CPU-Cache Cross-bar) interface. An inclusive L2 cache is distributed across all the tiles. The memory subsystem maintains cache coherence with a directory-based MESI coherence protocol. It adheres to the TSO memory consistency model used by the OpenSPARC T1. OpenPiton adds a new L1.5 cache to transduce CCX messages to the coherence protocol messages. Coherent messages between L1.5 caches and L2 caches communicate through three NoCs. A chip bridge connects the tile array to the chipset, through the upper-left tile, for serving memory and I/O requests. OpenPiton also includes an AXI4-Lite bridge that provides connectivity to a wide range of I/O devices by interfacing memory mapped I/O operations from the NoCs to AXI-Lite.

Three NoCs transport messages across the various message classes of the coherence protocol. NoC1 transports requests from L1.5 to the L2s; NoC2 transports responses and requests from the L2 to the cores and the memory controller respectively; NoC3 transports writebacks from the L1.5 and responses from the memory controller to the L2. To ensure deadlock-freedom across the message classes, the priority order among the NoCs is NoC3 > NoC2 > NoC1. This ensures that responses are always drained. The NoCs also maintain point-to-point ordering.

All NoCs use 64-bit bi-directional links. Each NoC uses wormhole routers without any virtual channels. The design essentially uses multiple physical networks instead of multiplexing multiple VCs over the same physical links. Dimension-ordered XY routing is used to avoid routing and protocol deadlocks. Each wormhole router takes one cycle when routing along the same dimension, and two cycles at turns. In the ASIC prototype, the NoC routers consume less than 3% of the entire chip area, which is dominated by the cores and caches.

8.3   INTEL XEON-PHI (2015)

Intel's Xeon Phi [321] line of processors is targeted for High Performance Computing (HPC) workloads and contain tens of cores. The first iteration, called Knights Corner was released in 2012. It is implemented in 22 nm and operates at 1–1.2 GHz. It contains 61 P54C (Pentium Pro) cores, interconnected by a bi-directional ring. The same ring is used in Intel's Xeon products. To get high-bandwidth and ring scalability, Intel uses 10 separate rings, 5 in each direction. The 5 rings are: one BL (64-byte for data), two AD (address), and two AK (acknowledgment and other coherence messages). All rings are of a different width, optimized for their traffic type. All packets are 1-flit wide, and each ring takes a cycle to deliver packets between ring stops. Apart from the cores, there are ring stops for eight memory controllers, PCIe controllers, and a few others for bookkeeping.

The second iteration, called Knights Landing was released in 2015. It is implemented in 14 nm and has 36 tiles, each with 2 silvermont (Atom) cores. A high level overview is shown in Figure 8.3. There are 38 physical tiles, of which 36 are active; the remaining 2 tiles are for yield recovery. Each tile comprises two cores, two vector processing units (VPUs) per core, a 1-Mbyte level-2 (L2) cache that is shared between the two cores, and a slice of the distributed directory. The NoC is a 6 × 6 mesh. There are four parallel meshes, each for delivering different types of traffic. The mesh can deliver greater than 700 GB/s of total aggregate bandwidth. There are no VCs within each mesh. The mesh is organized into rows and columns of "half" rings that fold upon themselves at the end points. In other words, the output link at the edge of an edge tile is connected to the input link at the same edge. All packets use YX routing: a packet first traverses the Y links to reach the right row, and then turns along the X to reach the destination. It takes

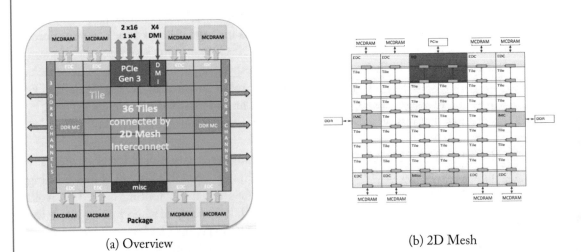

(a) Overview   (b) 2D Mesh

Figure 8.3: Intel Xeon Phi (knights landing): 6 × 6 mesh [321].

one cycle for every hop in the Y-direction, and two cycles for every hop in the X-direction. This is because the tile is wider on the X-direction than in the Y-direction. The mesh is derived from the rings in previous Xeon Phi generations and hence uses the same rotary rule for arbitration: traffic already on an X or Y ring always takes higher priority over traffic wanting to enter the ring from the injection or turning port.

The interconnect transports MESIF cache coherence traffic. The network can be operated in three cluster modes: all-to-all, quadrant, and sub-NUMA clustering. In all-to-all, addresses are hashed uniformly across all distributed directories, and there is no affinity between the tile, directory, or memory. In quadrant, the chip is divided into four virtual quadrants; the address is hashed to a directory in the same quadrant as the memory controller. In sub-NUMA clustering, each quadrant (cluster) is exposed as a separate NUMA domain to the OS; the system looks analogous to a 4-socket system and requires software to NUMA optimize to get benefit.

8.4   D E SHAW RESEARCH ANTON 2 (2014)

Anton 2 [338] from D E Shaw Research is a massively parallel special-purpose supercomputer designed to accelerate molecular dynamics (MD) simulations. Each Anton 2 node is an ASIC comprising 16 "Flex" and two "High-Throughput Interaction" compute subsystems. Hundreds of these nodes are connected as a 3-D torus to model 3-D MD systems, with two physical channels along each direction. The ASIC is implemented in 40 nm technology and runs at 1.5 GHz. Each ASIC has a 4×4 mesh NoC that serves two purposes: (a) connectivity between the compute systems and (b) switch for the torus channels. The mesh has two "skip channels" which are express links between the first and last router on the two rows connecting to the torus's X channels to mitigate latency. The topology is presented in Figure 8.4. All mesh links are 192-bit wide, capable of carrying 24-byte flits within a cycle. Each mesh channel has a bandwidth of 288 Gb/s, which is enough to route the torus bandwidth of 179 Gb/s, with substantial bandwidth left over for intra-node traffic.

Since both intra-node and inter-node traffic share the mesh, the routing algorithms on the mesh are optimized such that the NoC appears as a high bandwidth switch. The following direction-ordered route is reported to work best: Y-, X+, X- followed by Y+. The routers use 4 VCs each in the request and reply traffic classes (8 VCs in total) to avoid routing deadlocks. A VC is incremented every time a packet crosses a torus dimension, or crosses a dateline. Virtual-cut through (VCT) flow control is used. The network also supports table-based multicast to an arbitrary set of destinations. The routers have a 4-stage pipeline: route computation (RC), virtual channel allocation (VA), input switch arbitration (SA1) and output switch arbitration (SA2). Each takes about 0.7 ns in the design. The VC queues contribute to 46% of the entire router area.

In the Anton 2 network, the choice of a unified network for both intra-chip and inter-chip communication creates a fairness challenge, since each inter-chip hop is implemented as sequence of on-chip routing decisions, providing multiple opportunities for the introduction of

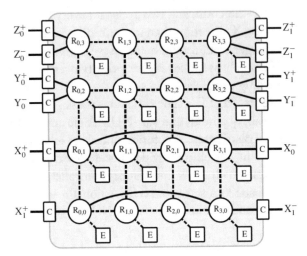

Figure 8.4: D E Shaw Research Anton 2: 4×4 mesh with skip channels [338].

unfairness. This is addressed by statically programming weights in the arbiters to provide service proportional to load, also known as EoS (equality of service). This static programming is possible since the class of MD applications running on the system is known.

8.5 MIT SCORPIO (2014)

The MIT SCORPIO chip [101] is a research prototype demonstrating snoopy coherence over an unordered mesh. It comprises 36 Freescale e200 Power Architecture cores, each with a private L1, connected by a 6×6 mesh. The core IPs have an AMBA AHB interface to a private L2, and the L2 has a AMBA ACE interface to a NoC. Two Cadence DDR2 memories attach to four unique routers along the chip edge, with the Cadence IP complying with the AMBA AXI interface, interfacing with the Cadence PHY to off-chip DIMM modules. All other IO connections go through an external FPGA board with the connectors for RS-232, Ethernet, and flash memory. The chip is fabricated in IBM 45 nm SOI and operates at 833 MHz. The chip layout is shown in Figure 8.5a.

The NoC is a 6×6 mesh. It implements sequential consistency by providing global ordering support, i.e., it guaranteees that snoop requests from cores are delivered to all destination cores in the exact same order. The SCORPIO NoC decouples message delivery and ordering by using 2 networks: a latency-bound bufferless broadcast network called a *notification network* to perform ordering and a *main network* to deliver the messages.

The notification network is a 36-bit mesh, where each bit corresponds to a source. For every coherence request injected on the main network, a one-hot encoding of the source node's ID (SID) is broadcast on the notification network. Broadcasts by multiple sources are merged

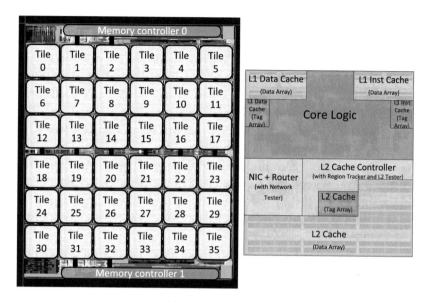

(a) Chip Overview

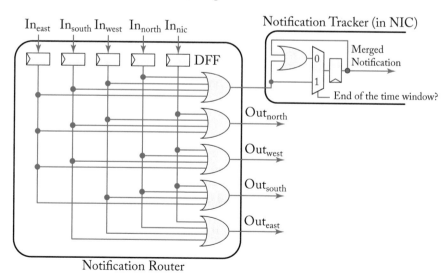

(b) Latency-bound Notification Network Router

Figure 8.5: MIT SCORPIO: 6 × 6 mesh [101].

by OR-ing the bit-vectors. The OR gates are hard-wired to implement an XY-tree routing algorithm. The contention-less design ensures a maximum latency bound for every broadcast. SCORPIO maintains synchronized time windows based on this bound; NICs send notifications at the beginning of the time window and process the notifications received at the end to the time window. Based on the notifications received within the time window, the corresponding requests are *ordered* in an ascending order of SIDs, thus enforcing the same global order. The actual requests can arrive over the main network in any order; the NICs deliver them to the coherence controllers in the global order. The NIC uses an *expected source ID (ESID)* register to keep track of the request it is waiting for. Once this request is received, it is sent to the controller and the ESID points to the next SID in the global order.

There is no flow control between the notification network switches since there is no buffering; however if the main network is slow, the notification network buffers at the destination NICs could fill up. This is handled by having an end-to-end flow control: a stop bit on the notification network is turned on by a NIC if it cannot receive any further notifications, and all senders wait until it is cleared.

The main network is a 6 × 6 mesh with broadcast support within the routers. All broadcasts use an XY-tree route. There are 4 VCs for the globally ordered broadcast requests, and 2 for the unicast responses. The former are 1-flit deep, and the latter 3-flit deep, and can each hold a full packet (i.e., virtual cut-through). At each NIC and router, one of the globally ordered VCs is *reserved* for the request whose SID matches the ESID that the NIC is waiting for. This ensures that at any time the request that the NICs are waiting for has reserved buffers all along the route and there are no deadlocks. The router in the main network has a 3-stage pipeline: buffer write + input VC arbitration, output port arbitration + VC selection, followed by switch traversal. Lookahead-bypass is also implemented and allows the pipeline to shrink to 1-cycle in the case of a successful bypass.

8.6 ORACLE SPARC T5 (2013)

The Oracle Sparc T5 [341] is a server chip optimized for database applications. The chip is implemented in 28 nm technology and runs at 3.6 GHz. It contains 16 cores, each 8-way multi-threaded, with a private L1 and a private L2. Each chip has a shared L3, which is 8-way banked. The NoC is a 8 × 9 crossbar, that interconnects 8 pairs of L2s to the 8 L3 banks and the IO controller (for off-chip accesses). The crossbar offers a bandwidth of 1 TB/s. Figure 8.6 presents the die photo of the chip. The crossbar area is about 1.5× the size of a L3 bank.

8.7 UNIVERSITY OF MICHIGAN SWIZZLE SWITCH (2012)

The Swizzle Switch [315] from the University of Michigan is an energy- and area-efficient crossbar topology that scales to high-radices. A 45 nm prototype is demonstrated for a radix-64 (i.e.,

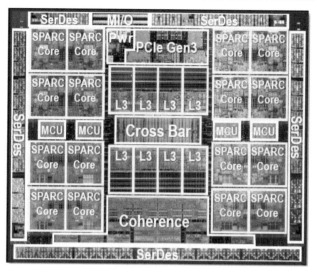

Figure 8.6: Oracle Sparc T5: 8 × 9 crossbar [341].

64 × 64) crossbar with 128-bit links. It operates at around 500 MHz delivering a throughput of 4.5 Tb/s. Figure 8.7 shows the chip floorplan.

The key idea in the Swizzle Switch is to re-use the data wires for arbitration, obviating the need for a separate control plane for arbitration which adds lot of area and delay penalties in crossbars due to the high fanout which in turn limits scalability. At each crosspoint, there is a vector of priority bits which specify which input ports this particular input port *inhibits*, i.e., has higher priority over. Each input port repurposes a particular bit of the horizontal input bus to assert a request, and a particular bit of the output bus to use as an inhibit line. Every output channel operates independently in two modes, *arbitration* and *data transmission*.

During the arbitration phase, all inhibit lines are pre-charged to 1. If an input channel has active data, it discharges the inhibit lines corresponding to the input ports it inhibits. For every output port, its highest priority input port wins arbitration and the result gets latched in a Granted Flip Flop to setup the connection for data transmission. During data transmission, the output buses are pre-charged to 1. At crosspoints where Granted Flip Flop is 1, the output remains charged or gets discharged based on the input. The Granted Flip Flop uses a thyristor-based sense amplifier to set the enabled latch, which only enables the discharge of the output bus for a short period of time, reducing the voltage swing on the output wire. This reduced swing coupled with the single-ended sense amplifier helps to increase the speed, reduce the crosstalk, and reduce the power consumption of the Swizzle-Switch.

The Swizzle Switch is proposed as a high-radix single-stage interconnect for a 64-core topology. It takes four cycles to use the Swizzle Switch: one cycle for the signals to reach the

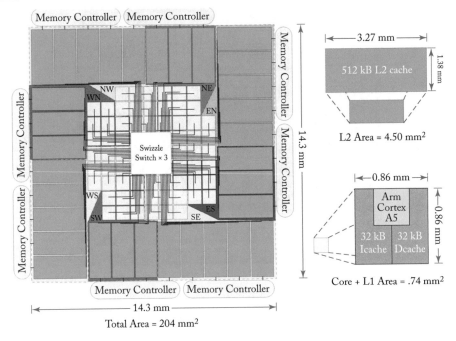

Figure 8.7: University of Michigan Swizzle Switch [315].

crossbar, one cycle for arbitration, one for data transmission, and one to reach the destination core.

8.8   MIT BROADCAST NOC (2012)

The Broadcast NoC [287] from MIT is a 16-node research prototype demonstrating a NoC with a single-cycle per-hop datapath optimized for broadcasts. The goal is to approach the latency, throughput and energy of an ideal broadcast fabric. The ideal metrics are derived and presented in the paper. The NoC is fed by on-chip traffic generators, which inject flits into the network according to a Bernoulli process of rate R, to a random, uniformly distributed destination for unicasts, and from a random, uniformly distributed source to all nodes for broadcasts. The injection rate, and the traffic mode (broadcast-only, unicast-only, or mix) is scanned in via input pins. The chip operates at 1 GHz and is implemented in 45 nm SOI CMOS. An overview of the chip is presented in Figure 8.8.

Each traffic generator connects to a NIC, which connects to a router. The routers are connected as a 4 × 4 mesh with 64-bit links. Broadcasts are routed over XY-trees and unicasts use XY. The design implements two message classes: requests and responses. Request packets, representing coherence requests and acknowledgments, are 1-flit wide, and could be broadcasts

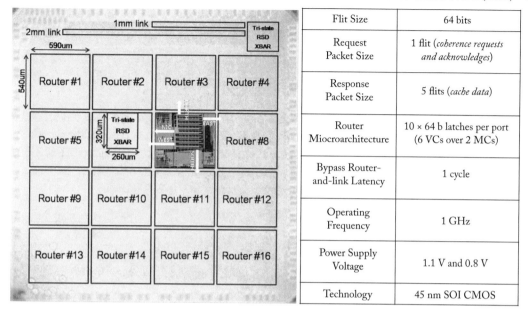

Flit Size	64 bits
Request Packet Size	1 flit (*coherence requests and acknowledges*)
Response Packet Size	5 flits (*cache data*)
Router Miocroarchitecture	10 × 64 b latches per port (6 VCs over 2 MCs)
Bypass Router-and-link Latency	1 cycle
Operating Frequency	1 GHz
Power Supply Voltage	1.1 V and 0.8 V
Technology	45 nm SOI CMOS

Figure 8.8: MIT Broadcast NoC: 4 × 4 mesh [287].

or unicasts. Response packets, representing cache lines, are 5-flits wide. There are 6 VCs in each router: four 1-flit deep for requests, and two 3-flit deep for responses.

The router implements two key features to approach the ideal latency, throughput and energy limits. The first is a multicasting crossbar, with low-swing links, which are designed to optimize both energy and latency. The links swing at 300 mV which leads to a 48.3% power reduction compared to an equivalent full-swing crossbar. It provides a single-cycle switch + link traversal (ST + LT) unlike conventional designs which spend a cycle each in the crossbar and the link. The crosspoints of the crossbar allow flits to get forked out of multiple output ports. The second feature of the router is the bypassing of pipeline stages to allow flits to arbitrate for multiple ports in one cycle and traverse the crossbar and links in the next, without having to stop and get buffered. This is implemented by sending 15 b lookaheads from the previous router to try and pre-arbitrate for one (or more) ports of the crossbar, one cycle before the actual flit. The arbiter is separable, the first stage (called mSA-I) arbitrates between input VCs at every input port, while the second stage (called mSA-II) arbitrates between input ports at every output port. The lookaheads have higher priority over local flits at each input port, and bypass mSA-I to directly enter mSA-II. If the lookahead wins arbitration for all ports, the incoming multicast flit is forked within the crossbar and not buffered at all. If the lookahead wins some or none of its output ports, the incoming flit is buffered and subsequently re-arbitrates for the remaining ports; partial grants are allowed. The regular router pipeline is three cycles: BW + mSA-I + VA in the first, mSA-II + lookahead traversal in the second, and ST+LT in the third. Successful arbitration by the lookaheads leads to the datapath becoming a single-cycle per-hop (ST+LT) low-swing traversal.

8.9 GEORGIA TECH 3D-MAPS (2012)

The 3D-MAPS (3D Massively Parallel Processor with Stacked Memory) chip from Georgia Tech [183] is a 2-tier 3-D chip with a 64-core processor on one layer, and SRAM on the other. An overview is shown in Figure 8.9. Each core communicates with its dedicated 4 KB SRAM block using face-to-face bond pads, which provide negligible data transfer delay between the core and the memory tiers. The maximum operating frequency is 277 MHz.

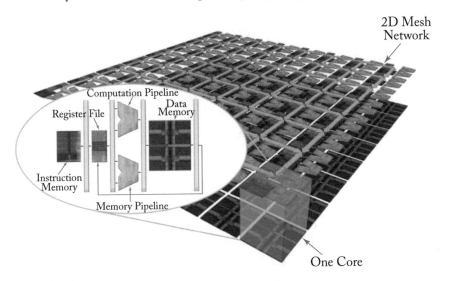

Figure 8.9: Georgia Tech 3D-MAPS: 3-D chip with 8×8 2-D mesh on logic tier [184].

Each core runs a modified version of MIPS, and implements an in-order dual-issue pipeline. A 2D mesh is used to connect the cores together, controlled by explicit communication and synchronization instructions. However there are no routers; explicit instructions are provided to move data generated by a core to its N, S, E, or W neighbor. The memory tier also has an 8×8 array of SRAM tiles, although these are not interconnected, and are private to each core.

8.10 KAIST MULTICAST NOC (2010)

The Multicast-NoC (MC-NoC) [191] from KAIST was developed as part of an application-specific chip for object recognition to be used in mobile robots. The SoC comprises 21 IP blocks:

a neural perception engine (NPE) for pre-processing, a task manager (TM) for scheduling, 16 SIMD processor units (SPU), a decision processor (DP) for post-processing, and 2 external interfaces (EXT). All IPs operate at 200 MHz. The IPs connect via a NoC that operates at twice the IP frequency, i.e., at 400 MHz. The chip was fabricated in 0.13 um CMOS.

The NoC topology is a *hierarchical star-ring (HS-R)* shown in Figure 8.10. The SPUs are grouped in clusters of 4. The clusters are connected together by a ring. A 7 × 7 local switch within each cluster connects the 4 SPUs and the two directions of the ring to a system network switch. The 9 × 10 system network switch connects together the NPE, TM, DP, EXTs, and the four clusters. There are two ports into the DP; one is dedicated for aggregating N-to-1 notification packets from the SPUs. The combined topology of the MC-NoC provides 118.4 GB/s total bandwidth with a maximum of 3 switch traversals for any packet.

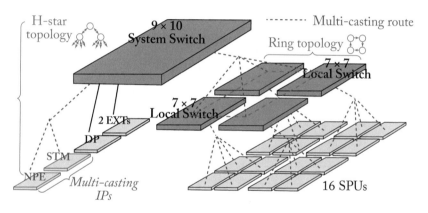

Figure 8.10: KAIST MC-NOC: Hierarchical star-ring [191].

Source-routing is used for both unicasts and multicasts. The header flits carry 16 bits of routing information that is used to specify the route for unicasts, and the destination SPU set for multicasts. The header also carries a 4-bit burst length for data bursts of up to 8 flits per packet, and a 2-bit priority for quality-of-service.

There are no VCs; wormhole flow control is used. Each router has a 4-stage pipeline. In the first stage, an incoming flit's header is parsed and it is buffered in a 8-depth FIFO that manages synchronization for heterogeneous clock domains within the IPs and the NoC. In the second stage, active input ports send request signals to each output port arbiter. The arbiters perform round-robin scheduling according to the priority levels of the requests. In the third stage, the grants are received. For multicasts, a grant checker is used to check if all requesting output ports were granted or not. If they were, the flit is dequeued and broadcast out of the crossbar in the fourth stage. If not, the flit retries next cycle for all ports as partial grants are not allowed. A variable strength driver is employed at every input port of the crossbar to provide sufficient driving strength for multicasting.

8.11 INTEL SINGLE-CHIP CLOUD (2009)

The Intel SCC [160] is a 48-core research prototype to study many-core architectures and their programmability. All 48 IA-cores boot Linux simultaneously. The chip is implemented in 45 nm CMOS and operates at 2 GHz. There is no hardware cache coherence; instead software maintains coherence using message passing protocols such as MPI and OpenMP. The SCC has 24 tiles, each housing 2 cores and a private L1 and L2 per core. The tiles are connected by a 6×4 mesh NoC offering a bisection bandwidth of 256 GB/s. Figure 8.11 shows an overview of the chip.

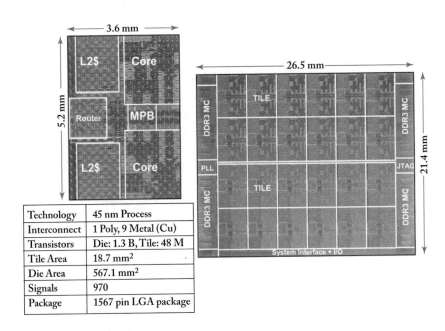

Technology	45 nm Process
Interconnect	1 Poly, 9 Metal (Cu)
Transistors	Die: 1.3 B, Tile: 48 M
Tile Area	18.7 mm^2
Die Area	567.1 mm^2
Signals	970
Package	1567 pin LGA package

Figure 8.11: Intel Single-Chip Cloud: 6×4 mesh [160].

The routers have 5 ports with each input port housing five 24-entry queues, a route pre-computation unit, and a virtual-channel (VC) allocator. Route pre-computation for the outport of the next router is done on queued packets. The links are 16 B wide, with 2 B sidebands. Sideband signals are used to transmit control information. XY dimension-ordered routing is enforced. There are a total of 8 VCs, two reserved for the request and response message classes and the rest in a free pool. Credit-based flow control is used between the routers. Input port and output port arbitrations are done concurrently using a wrapped wave front arbiter. The router uses virtual cut-through flow control and performs crossbar switch allocation in a single clock cycle on a packet granularity. The router has a 3-cycle pipeline: Input Arbitration, Route Pre-Compute + Switch Arbitration, followed by VC allocation. This is followed by a 1-cycle link traversal to the next router. A packet consists of a single flit or multiple flits (up to three) with header, body and tail flits.

The die is divided into 8 voltage islands, and 28 frequency islands to allow the V/F of each island to be independently modulated by software. The complete 2D mesh is part of one V/F domain. The NoC contributes to 5% and 10% of the total power at low-power (0.7 V, Cores - 125 MHz, Mesh - 250 MHz) and high-power (1.14 V, Cores - 1 GHz, Mesh - 2 GHz) operation, respectively.

8.12 UC DAVIS ASAP (2009)

The Asynchronous Array of simple Processors (AsAP) [339] from the University of California at Davis is a programmable array of processors and accelerators. The first generation chip which is in 0.18 um contains 36 cores, and operates at over 610 MHz at 2 V. The second generation chip which is fabricated in ST Microelectronics 65 nm low-leakage CMOS process comprises 164 RISC processors, three fixed-function accelerators (FFT, Viterbi decoder, and Motion Estimation), and three 16KB shared memory modules. A high-level overview of the chip is shown in Figure 8.12a. At 1.3 V, the programmable processors can operate up to 1.2 GHz. The configurable FFT and Viterbi processors can run up to 866 MHz and 894 MHz, respectively.

The cores are connected to their nearest neighbors via two sets of links, forming two separate 2-D mesh networks. Applications are mapped on to the array, and *circuit-switched* paths are created between each communicating pair of cores. Having two meshes eases the mapping job for programmers. The clock is distributed in a globally asynchronous, locally synchronous (GALS) manner by sending the source clock along with the data and using it to latch the data at the destination FIFO. This obviates the need for a global clock distribution tree over the chip. The destination core uses its own internal clock to read data out of the FIFO.

The connection between two cores can pass through multiple intermediate switches depending on the mapping. This interconnection is established by configuring the multiplexers in the intermediate switches prior to runtime which fixes this communication path; thus, this static circuit-switched interconnect is guaranteed to be independent and never shared. As long as the destination processor's FIFO is not full, a very high throughput of one data word per cycle can be sustained.

The number of "cycles" taken for data communication depends on the distance between the communicating cores. Since data is not latched at intermediate switches, the entire path is a repeated link connecting the source processor's FIFO with the destination processor's FIFO with pre-configured 4:1 muxes at each "switch." Multiple switches can thus be traversed within a single clock cycle. A typical communication path is shown in Figure 8.12b. The latency observed (in simulation) is less than 2.5 cycles at 90 nm and less than 1.7 cycles at 22 nm at the peak operating frequency for each technology, regardless of distance.

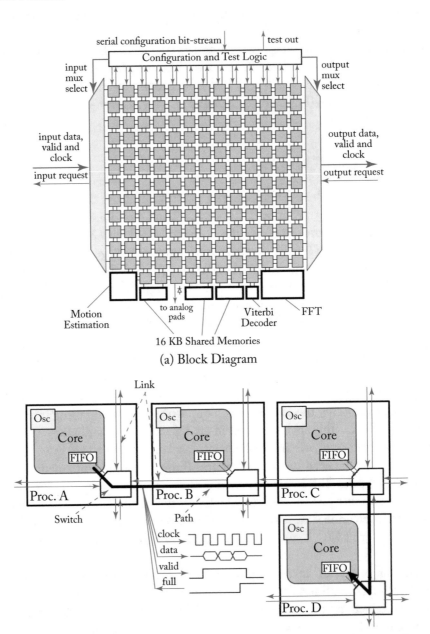

(a) Block Diagram

(b) Circuit-switched path between two processors through intermediate switches

Figure 8.12: UC Davis AsAP: 13×13 mesh [339].

8.13   TILERA TILEPRO64 (2008)

The Tilera TILE*Pro*64 [356] is a multicore SoC targeting embedded applications across networking and digital multimedia. It comprises 64 tiles connected by multiple independent 8 × 8 meshes collectively called *iMesh*, as shown in Figure 8.13. Each mesh consists of 32-bit unidirectional links. Each tile features a processor engine, cache engine, and switch engine. The core is a 64-bit VLIW engine, capable of running SMP Linux, with a private L1-D, L1-I, and L2 cache. The cache engine is a distributed shared L3 slice. The switch engine houses the routers for the various networks.

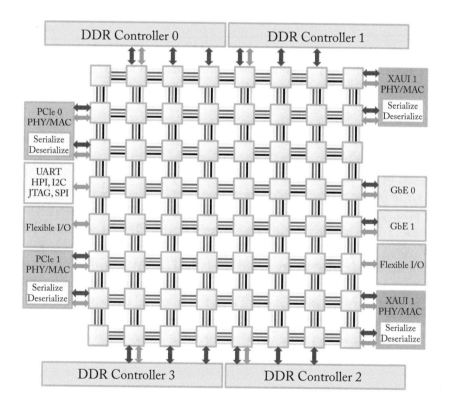

Figure 8.13: Tilera TILE*Pro*64: 6 8 × 8 mesh [356].

At the chip frequency of 1 GHz, iMesh provides a bisection bandwidth of 320 GB/s. Actual realizable throughput depends on the traffic and how it is managed and balanced by the flow control and routing protocols.

The iMesh has sophisticated network interfaces supporting both shared memory and message passing paradigms. There are six physical networks: Memory Dynamic Network (MDN), Tile Dynamic Network (TDN), User Dynamic Network (UDN), Static Network (STN), and

I/O Dynamic Network (IDN), and in Validation Dynamic Network (VDN). Traffic is statically divided across the six meshes. The caches and memory controllers are connected to the MDN and TDN with inter-tile shared memory cache transfers going through the TDN and responses going through the MDN, providing system-level deadlock freedom through two separate physical networks. The UDN supports user-level messaging, so threads can communicate through message passing in addition to the cache coherent shared memory. Upon message arrivals, user-level interrupts are issued for fast notification. Message queues can be virtualized into off-chip DRAM in case of buffer overflows in the NIC. The STN is used for routing large streaming data. I/O and system messages use the IDN. The TILE64 contains these five networks; the VDN was introduced in the TILE*Pro*64 for invalidation traffic to accelerate cache coherence.

The five dynamic networks (UDN, IDN, MDN, TDN, and VDN) use the dimension-ordered routing algorithm, with the destination address encoded in X-Y coordinates in the header. The static network (STN) allows the routing decision to be pre-set. This is achieved through circuit switching: a setup packet first reserves a specific route, the subsequent message then follows this route to the destination.

The dynamic networks use simple wormhole flow control without virtual channels to lower the complexity of the routers, trading off the lower bandwidth of wormhole flow control by spreading traffic over multiple networks. Credit-based buffer management is used. The static network uses circuit switching to enable the software to pre-set arbitrary routes while enabling fast delivery for the subsequent data transfer; the setup delay is amortized over long messages.

Buffer management in each network is varied. On the MDN, a conservative end-to-end approach is used, where in every node communicating with DRAM is allocated a slot at the memory controller. This guarantees that traffic on the MDN is always drained, without causing any congestion. Acknowledgments are issued when the DRAM controller processes a request. The storage at the memory controller is sized to cover the acknowledgment latency and allow multiple in-flight memory requests. On the TDN, the link-level flow control is used. As long as the MDN drains (due to the end-to-end flow control), the TDN can make forward progress. The IDN and UDN are software accessible and implement mechanisms to drain into the DRAM, and refill, to avoid deadlocks. In addition, the IDN utilized pre-allocated buffering with explicit acknowledgments when communicating with I/O devices. The UDN can employ multiple end-to-end buffer management schemes depending on the programming model.

The iMesh' wormhole networks have a single-stage router pipeline during straight portions of the route, and an additional route calculation stage when turning. Only a single buffer queue is needed at each of the five router ports, since no VCs are used. Only three flit buffers are used per port, just sufficient to cover the buffer turnaround time. This emphasis on simple routers results in a low area overhead of just 5.5% of the tile footprint.

8.14 ST MICROELECTRONICS STNOC (2008)

ST Microelectronic's STNoC [82] aims to provide a programmable on-chip communication platform on top of a simple network for heterogeneous multicore platforms. It encapsulates support for communication and synchronization primitives and low-level platform services within what it calls the Interconnect Processing Unit (IPU). Examples of communication primitives are send, receive, read and write, while synchronization primitives are test-and-set and compare-and-swap. The aim is to have a library of different IPUs that support specific primitives so MPSoC designers can select the ones that are compatible with their IP blocks. For instance, IP blocks that interface with the old STBus require read-modify-write primitives that will be mapped to appropriate IPUs. Currently STNoC supports two widely used SoC bus standards fully: the STBus and AMBA AXI [27], and plans to add IPUs for other programming models and standards.

The STNoC proposes a novel pseudo-regular topology, the Spidergon, that can be readily tailored depending on the actual application traffic characteristics, which are known a priori. Figure 8.14 sketches several variants of spidergons. Figure 8.14a shows a 6-node spidergon that can have more links added to cater to higher bandwidth needs (Figure 8.14b). Figure 8.14c shows a maximally connected 12-node spidergon, where most links can be trimmed off when they are not needed (Figure 8.14d). The pseudo-regularity in STNoC permits the use of identical degree three router nodes across the entire range of spidergon topologies, which simplifies design and makes it easier for a synthesis algorithm to arrive at the optimal topology. A regular layout is also possible, as Figure 8.14e illustrates.

The STNoC can be routed using regular routing algorithms that are identical at each node, leveraging the ring-like topology of the spidergon. For instance, the Across-First routing algorithm sends packets along the shortest paths, using the long across links that connect non-adjacent nodes in STNoC only when that gives the shortest paths, and only as the first hop. For instance, in Figure 8.14f, when going from Node 0–4, packets will be routed from Node 0–3 on the long across link, then from 3–4 on the short link, leading to a 2-hop route. Note though that here, clearly, link length differs significantly and needs to be taken into account. Despite a low hop count, long link traversals cycles may increase the packet latency for Across-First routing. The Across-First routing algorithm is not deadlock-free, relying on the flow control protocol to ensure deadlock freedom instead.

STNoC routing is implemented through source routing, encoding just the across link turn and the destination ejection, since local links between adjacent rings are default routes. The Across-First algorithm can be implemented within the network interface controller either using routing tables or combinational logic.

STNoC uses wormhole flow control, supporting flit sizes ranging from 16–512 bits depending on the bandwidth requirements of the application. Virtual channels are used to break deadlocks, using a dateline approach similar to what has been discussed, with the variant that nodes that do not route past the dateline need not be constrained to a specific VC, but can in-

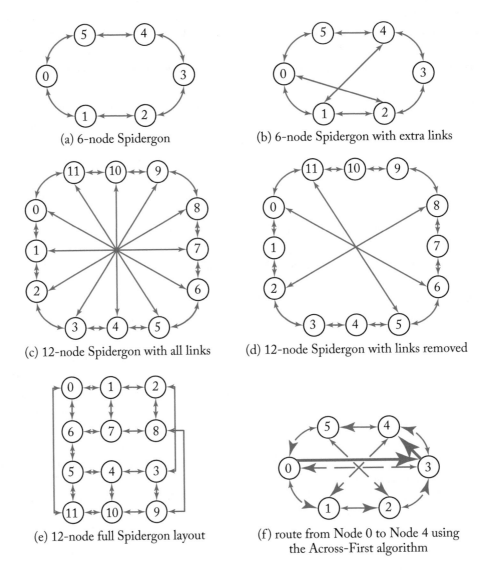

(a) 6-node Spidergon

(b) 6-node Spidergon with extra links

(c) 12-node Spidergon with all links

(d) 12-node Spidergon with links removed

(e) 12-node full Spidergon layout

(f) route from Node 0 to Node 4 using the Across-First algorithm

Figure 8.14: STNoC: Various Spidergon topologies.

stead use all VCs. Buffer backpressure is tracked using credits. Actual flit size, number of buffers, and virtual channels are determined for each application-specific design through design space exploration.

Since the STNoC targets MPSoCs, it supports a range of router pipelines and microarchitectures. Here we summarize the key attributes that are common across all STNoC routers. As MPSoCs do not require GHz frequencies, STNoC supports up to 1 GHz in 65 nm ST technology. Since the degree is three, the routers are four-ported, to left, right, across links as well as the NIC to IP blocks. Low-load bypassing is used to lower the latency. Buffers are placed at input ports, statically allocated to each VC, with the option of also adding output buffer queues at each output port to further relieve head-of-line blocking. Round-robin, separable allocators are used. Extra ports or links that are trimmed off are not instantiated at design time. The crossbar switch is fully synthesized.

8.15   INTEL TERAFLOPS (2007)

The Intel TeraFLOPS chip [158] was one of the first many-core industry prototypes. It was built in 65 nm, and could run at up to 5 GHz. It houses 80 single-precision, floating point cores. The chip details are shown in Figure 8.15a.

The processing engine (PE) in each tile contains two independent, fully pipelined, single-precision floating-point multiply-accumulator (FPMAC) units; 3 KB of single-cycle instruction memory (IMEM); and 2 KB of data memory (DMEM). A 96-bit very long instruction word (VLIW) encodes up to eight operations per cycle. Applications are hand-mapped over the PEs, with communication managed via explicit message passing send/receive instructions in the ISA. Any tile can send/receive to any other tile, and send/receive instructions have latencies of two cycles (within a tile's pipeline), and five cycles of router pipeline along with a cycle of link propagation for each hop. The 2-cycle latency of send/receive instructions is the same as that of local load/store instructions. In addition, there are sleep/wakeup instructions to allow software to put entire router ports to sleep for power management. These instructions trigger sleep/wakeup bits in the packet header to be set, which can turn on/off 10 sleep regions in each router as they traverse the network.

The topology is a 8×10 mesh, with each channel composed of two 38-bit unidirectional links. It runs at an aggressive clock of 5 GHz on a 65 nm process. This design gives it a bisection bandwidth of 380 GB/s or 320 GB/s of actual data bisection bandwidth, since 32 out of the 38 bits of a flit are data bits; the remaining 6 bits are used for sideband. Each tile has a mesochronous interface (MSINT) which allows for scalable clock phase-insensitive communication across tiles and synchronous operation within each tile.

The network uses source table-based routing, with each hop encoded as a 3-bit field corresponding to the 5 possible output ports a packet can take at each router. The packet format supports up to 10 hops (30 bits for route information), with a chained header bit that when set, indicates that routing information for more hops can be accessed in the packet data. Source

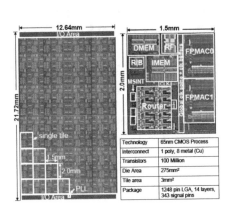

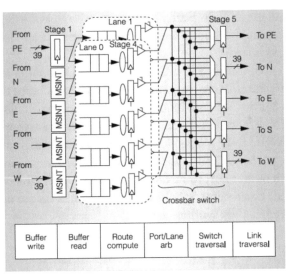

(a) Chip Overview

(b) Router Microarchitecture and Pipeline

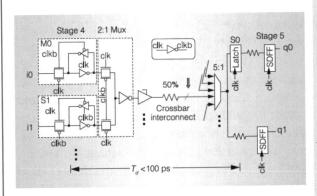

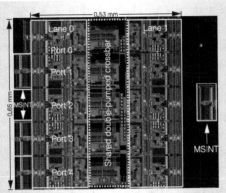

(c) Double-pumped Crossbar Circuit

(d) Router Layout

Figure 8.15: Intel TeraFLOPS: 8 × 10 mesh [158].

routing enables the use of many possible oblivious or deterministic routing algorithms or even adaptive routing algorithms that chooses a route based on network congestion information at injection point. This enables the TeraFLOPS to tailor routes to specific applications.

The TeraFLOPS has a minimum packet size of two flits (38-bit flits comprised of 6 bits of control data and 32 bits of data), with no limits placed on the maximum packet size by the router architecture. The network uses wormhole flow control with two virtual channels (called "lanes"), although the virtual channels are used only to avoid system-level deadlock, and not for flow control, as packets are pinned to a VC throughout their network traversal. This simplifies the router design since no VC allocation needs to be done at each hop. Buffer backpressure is maintained using on/off signaling, with software programmable thresholds.

The high 5 GHz (15 FO4s) frequency of the TeraFLOPS chip mandates the use of aggressively pipelined routers. The router uses a five-stage pipeline: buffer write, route computation (extracting the desired output port from the header), two separable stages of switch allocation, and switch traversal. The pipeline is shown in Figure 8.15b. It should be noted that single hop delay is still just 1 ns. Each port has two input queues, one for each VC, that are each 16 flits deep. The switch allocator is separable in order to be pipelineable, implemented as a 5–1 port arbiter followed by a 2–1 lane arbiter. The first stage of arbitration within a particular lane essentially binds one input port to an output port, for the entire duration of the packet, opting for router simplicity over the flit-level interleaving of multiple VCs. Hence, the VCs are not leveraged for bandwidth, but serve only deadlock avoidance purposes.

The crossbar switch is custom-designed, using bit-interleaving, or double pumping, with alternate bits sent on different phases of a clock, reducing the crossbar area by 50%. The crossbar is fully nonblocking, with a total bandwidth of 100 Gbytes/s. The crossbar circuit is shown in Figure 8.15c. The layout of the router is custom, with the crossbar at the center, and the queues, control, and arbiters for each VC (lane) on either side.

The maximum frequency of the router ranges from 1.7 GHz at 0.75 V to 5.1 GHz at 1.2 V. The corresponding measured on-chip network power per tile with all router ports active ranges from 98–924 mW, consuming 39% of the tile power. Clock gating and sleep transistors at every port help reduce dynamic and leakage power respectively, and can lower the total power to 126 mW, a 7.3× reduction.

8.16   IBM CELL (2005)

The IBM Cell Broadband engine [195] is a multicore chip that drives Sony's PlayStation 3. It was built in 65 nm, and runs at 3.2 GHz It includes one POWER Processing Element (PPE) and eight Synergistic Processing Elements (SPEs). An Element Interconnect Bus (EIB) interfaces with the the PPE, the SPEs, the memory controller, and two I/O interfaces in and out of the chip. A high-level overview is presented in Figure 8.16.

The EIB consists of four unidirectional rings, two in each direction, and operates at half the processor-clock speed, i.e., at 1.6 GHz. Each ring can simultaneously send and receive 16 bytes of data every bus cycle. The EIB's maximum data bandwidth is limited by the rate at which addresses are snooped across all units in the system, which is one address per bus cycle. Each snooped address request can potentially transfer up to 128 bytes, so in a 3.2 GHz Cell processor, the theoretical peak data bandwidth on the EIB is 128 bytes × 1.6 GHz = 204.8 Gbytes/s.

As the Cell interfaces with the EIB through DMA bus transactions, the unit of communications is large DMA transfers in bursts, with flow control semantics of buses rather than packetized networks. Resource allocation (access to rings) is guarded by the ring arbiter, with

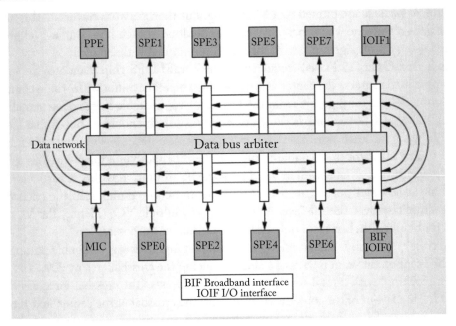

Figure 8.16: IBM Cell: 4 rings [195].

highest priority given to the memory controller so requestors will not be stalled on read data. Other elements on the EIB have equal priority and are served in a round-robin manner.

The IBM Cell uses explicit message passing as opposed to a shared-memory paradigm. It is designed to preserve DMA over split-transaction bus semantics, so snoopy coherent transfers can be supported atop the four unidirectional rings. In addition to the 4 rings, these 12 elements interface to an address-and-command bus that handles bus requests and coherence requests. The rings are accessed in a bus-like manner, with a sending phase where the source element initiates a transaction (e.g., issues a DMA), a command phase through the address-and-command bus where the destination element is informed about this impending transaction, then the data phase where access to rings is arbitrated and if access is granted, data are actually sent from source to destination. Finally, the receiving phase moves data from the NIC (called the Bus Interface Controller (BIC)) to the actual local or main memory or I/O.

8.17   CONCLUSION

As evident by the case studies in this chapter, there has been a significant uptick in commercial design and research prototypes featuring on-chip networks. Although meshes remain the most common topology, rings, and crossbars continue to be optimized. Dimension-ordered routing is widely favored across the designs studied due to its simplicity and deadlock freedom. Wider variation is seen in both flow control methods and router pipeline stages. Finally, these commercial design and prototypes highlight the importance of considering the entire system; supporting cache coherence protocols, message passing, and broadcasting/multicasting features in many of the designs. Although common attributes have emerged as the field of on-chip networks matures, we anticipate exciting new research in all aspect of on-chip networks to drive the field in the next decade.

CHAPTER 9

Conclusions

The study of on-chip networks is a relatively new research field. Conference papers addressing them began appearing only in the late 1990s and on-chip networks have only recently began appearing in products in sophisticated forms. In this concluding chapter, we reflect on emerging research challenges in this field, surveying the state of the art and summarizing several key opportunities.

9.1 BEYOND CONVENTIONAL INTERCONNECTS

On-chip networks are inherently comprised of underlying devices that drive network router logic and interconnects that enable the transfer and switching of control and data signals through the network. With communications as its main functionality, on-chip networks are naturally driven by research advancements in interconnects.

Throughout this lecture, we intentionally omitted interconnect design, focusing on router architecture and design instead. This is because to date, all industry chips with on-chip networks used conventional repeated wires at full voltage swing. Conventionally, upper metal levels such as M5 or 6 are used for the longer links of on-chip networks, and repeaters are automatically inserted during back-end layout, appropriately trading off delay with energy [274]. Recent years have seen significant progress in alternative interconnects for on-chip networks which we will briefly survey and discuss below.

Low-power interconnect I/O circuit design. To drive on-chip network power down, power-efficient link designs can be used to lower transmission energy. Researchers have explored leveraging off-chip link I/O techniques [92], such as low voltage swing signaling, equalization and differential signaling for on-chip transmission lines [60, 61, 165, 171, 181, 219, 229, 288]. Designers have to carefully trade off the benefits in energy, delay or bandwidth with the much tighter area/power budgets for on-die networks. For instance, while low swing signaling has been demonstrated within on-chip network research chip prototypes [219, 288], the reliability of such links still needs to be rigorously evaluated in advanced technology nodes where supply voltages decrease and the headroom for low swing signaling narrows. These sophisticated interconnects have also prompted novel architectures that leverage their unique characteristics of fast cross-chip delay, or a global shared medium interconnect [56, 57, 205, 270].

A key impediment to the adoption of power-efficient link circuits within NoCs lies in the disconnect between these custom designed low-power transmitter/receiver (TX/RX) cells and the VLSI CAD flow that is essential for handling the complexity of the many-core NoC chips.

Recent research has targeted this problem in CAD for NoCs, proposing toolchains that enable the embedding of custom-designed TX/RX cells automatically within the commercial CAD flow [65], paving the way for such advanced interconnects to be part of mainstream NoCs.

Photonics NoCs. Optical fibers have already effectively replaced electronic cables for inter-chassis interconnections within data centers, and optical backplanes are emerging as a viable alternative between racks of a chassis [121]. Recently, researchers have also proposed that photonics offer promise for shorter interconnects, from off-die I/O to DRAM, to even on-die interconnects for on-chip networks. Miller [248] projected that electrical on-chip networks will be unable to achieve the bandwidth demands and tight power envelope of future chips, detailing how the physical characteristics of optics and the progress in optical device research can potentially solve that. Research into photonics materials and devices has advanced significantly in the last decade [46, 228, 232, 249, 329], demonstrating increasingly robust photonics devices with good power-performance on silicon-compatible materials. Key progress lies in increasing compatibility with commercial foundries, a major stumbling block in the incorporation of photonics into many-core NoC chips. A working microprocessor chip with opto-electronic off-die processor-memory links to DRAM was demonstrated in an IBM 45 nm SOI CMOS process with an external laser, showing that working photonics devices can be achieved in a commercial silicon process [329]. On-die photonics may be within reach, if the energy per bit can be driven further down to compete with on-die copper interconnects.

While photonics device and materials research has to be the driving force behind research into on-chip photonics, multi-disciplinary research is needed into a whole range of areas, spanning from circuits and CAD to architecture. In materials and device research, realizing on-die photonics will prompt further investigation into on-die lasers and light sources, processed at large scale in a manner compatible with commercial silicon processes. Electrical circuits that interface to the photonics devices need to not only push the power envelope down to few fJ/bit but also deliver that in the face of emerging photonics devices with significant variance in device performance due to limited characterization as compared to mature CMOS devices. A photonics CAD flow will have to be built from ground up to enable VLSI-scale design of photonics devices. Several multidisciplinary groups have already made significant progress: on-die lasers and LEDs have been recently demonstrated [376], low-power opto-electronic link circuits have been fabricated [330, 373], photonics CAD flow that is compatible with existing CAD tools have been researched [55, 78, 79, 104, 150, 346], and alternative optical on-chip network architectures inspired by photonics advances in light sources, modulators, detectors and even photonic switches have been looked into [77, 172, 194, 197, 198, 212, 225, 279, 280, 316, 345, 379].

Wireless NoCs. Millimeter (mm)-wave wireless networks-on-chip are being researched for providing low-latency high-bandwidth long-distance on-chip communication [103, 192]. Wireless NoCs also mitigate layout challenges for long electrical links. A key benefit of these links is their inherent broadcast capability, which can be used for efficient coherence [5] and synchronization [4]. Research into transceiver and antenna design for wireless NoCs and a demonstration of a robust functioning prototype is a key for this emerging technology to be used in a more widespread manner.

Off-die I/Os. With off-chip I/O being constrained by the number of pins and link bandwidth, on-chip network researchers have also explored alternatives to off-die I/O and the implications on on-chip network architecture and design. 3-D stacking has matured rapidly in the past years and provides a viable option for very high bandwidth off-chip I/O. 3-D stacking, leading not only to a substantial increase in the demand for on-chip bandwidth, but also impacting the design of on-chip networks. Off-die memory controllers that conventionally have to be placed on the borders of chips need no longer be constrained, affecting on-die network traffic flows spatially. The large bandwidth enabled by 3-D stacking naturally affects the temporal flow of traffic to the on-chip network, affecting topology significantly, prompting several research works in architecture and circuits forums [112, 175, 176, 188, 223, 286, 337, 365, 366]. The 3D-Maps processor chip from Georgia Tech discussed in our earlier chapter on case studies showcases the realization of a 3-D many-core chip and the impact of 3-D stacking on on-chip network design. Wireless inductive-coupling links for connecting multiple die on package are also being explored [332].

9.2   RESILIENT ON-CHIP NETWORKS

As technology scales toward deep sub-micron lithographies, an on-chip network that is architected and designed assuming perfect, zero-fault fabrication and operation will no longer suffice. This is aggravated by the system overlaid atop the on-chip network typically assuming always-correct communication. For instance, a cache-coherent shared-memory CMP is architected assuming that every message will be delivered correctly to the designated destination, with no packets dropped mid way through transmission. Similarly, a MPSoC assumes that every transaction through the on-chip network is successfully completed.

Shrinking gate lengths will lead to manufacturing defects and variability, with the large die sizes of many-core chips making post-fabrication faults highly likely [32]. When combined with errors that are likely to occur during chip operation, such as soft errors and wearout errors, it will be critical to design on-chip networks that are resilient and continue correct operation in the face of many faults/errors.

Fault-tolerant routing has been investigated substantially in the past, in the domain of clusters of workstations [110, 294, 374] and large-scale, multi-chassis, multi-computers [90], and can be leveraged. In on-chip networks, resilient mechanisms have to be designed under very tight power and area constraints, and yet work in the face of many faults. In recent years, substantial research has been done in this area [300], from resilient routing that reconfigures around faults [105, 119, 120, 122, 196, 282], to fault-tolerant router microarchitectures with built-in redundancy [80]. There have also been research into resilient link designs that trade off link power and reliability [363], and NoC fault modeling tools for characterizing how advanced processes can impact NoC operation [19]. The testing community has also looked extensively into NoC testing in the face of faults [140, 284, 296].

9.3   NOCS AS INTERCONNECTS WITHIN FPGAS

Communication is becoming a bottleneck given the large and increasing size of FPGAs. Modern FPGAs integrate fixed function IPs such as DSPs, processors, memory and high speed I/O with a large reconfigurable fabric. The fine-grained bit-level control over individual signals within the FPGA fabric creates significant design challenges for engineers as they must orchestrate communication and iterate over their design to ensure timing closure. To ease these burdens, recent work has proposed embedding a NoC within the FPGA fabric. The NoC provides the high performance communication substrate while allowing designers to focus on implementation of computational blocks.

Given the flexible and reconfigurable nature of an FPGA, one of the key questions regarding the implementation of a NoC for FPGAs is whether that NoC should be soft, i.e., built out of the FPGA fabric itself or hardened, i.e., implemented in dedicated silicon alongside the FPGA fabric. Soft NoCs provide improved design modularity and better scalability without requiring large scale changes to the FPGA hardware itself [106, 177, 281, 305]. Such research looks at how to efficiently implement a NoC out of a reconfigurable fabric. However, soft NoCs can suffer from large area overheads and low operating frequencies. This motivates the consideration of hardened NoC architectures for FPGAs [6, 7, 8, 9, 76, 130, 133, 239]. These hardened NoCs can deliver higher performance and greater efficiency while only requiring a small fraction of silicon area. Beyond the architecture of the NoC itself, their use in FPGA require their integration into the CAD flow so that designers can easily make use of the bandwidth and performance that they offer.

9.4   NOCS IN ACCELERATOR-RICH HETEROGENEOUS SOCS

As performance and energy benefits from technology scaling have started to diminish, there is an emergence of a new class of on-chip SoCs: heterogeneous accelerator-based architectures. CPU-GPU architectures were the first to emerge in this class of devices. There has been relatively little work exploring NoC architectures for GPUs. When considering these types of systems, one has to be cognizant of the different memory access patterns seen inside GPUs, most notably the asymmetry in requests (many-to-few from L1 to L2 banks) and responses (few-to-many from L2 banks to L1s). Tailoring the network to the specific memory traffic behavior can lead to more efficient designs [35, 372, 378]. Additional work exploring the integration of emerging technologies such as photonics NoCs inside GPU NoCs is also being explored [136, 377].

Moving beyond GPU-based accelerators, we anticipate a wider array of accelerator-rich SoC architectures. Such SoCs fall in between the traditional application-specific MPSoCs and fully general-purpose homogeneous CMPs. The design of NoCs connecting heterogeneous IPs, each with unique latency/bandwidth requirements across time, is an important research question. The design of the NoC *within* an accelerator is also an open question. Accelerators are built using tiny compute elements, and often operate in dataflow style, where the delivery of a piece of data triggers some action in the compute element. This makes the NoC critical to the throughput of such accelerators. Recent accelerator prototypes for deep neural networks [72, 107], spiking neural networks [21], databases [364], and graph processing [17] all use customized NoCs. NoCs for accelerator-based systems is expected to become an active area of research going forward.

9.5   ON-CHIP NETWORKS CONFERENCES

Research into on-chip networks is appearing in many venues spanning several disciplines. Table 9.1 highlights some major conferences in Architecture, CAD, VLSI, and NoCs that publish innovative research into on-chip networks. This list is not exhaustive; in addition to major conferences, recent years have seen workshops specifically focused on on-chip networks at many of these conferences.

Table 9.1: On-chip network conferences

Field	Conference
Architecture	International Symposium on Computer Architecture (ISCA)
	International Symposium on Microarchitecture (MICRO)
	International Symposium on High Performance Computer Architecture (HPCA)
	International Conference on Parallel Architectures and Compilation Techniques (PACT)
	International Conference on Architectural Support for Programming Languages and Operating Systems (ASPLOS)
CAD	International Conference on Computer-Aided Design (ICCAD)
	Design Automation Conference (DAC)
	Design Automation and Test in Europe (DATE)
VLSI	International Conference on VLSI (VLSI)
	International Solid State Circuits Conference (ISSCC)
Network on Chip	International Network on Chip Symposium (NOCS)

9.6   BIBLIOGRAPHIC NOTES

Finally, we refer the reader to other summary and overview papers [44, 50, 151, 238, 267, 276] to help guide them in further study of on-chip networks.

References

[1] Pablo Abad, Pablo Prieto, Lucia G Menezo, Valentin Puente, and José-Ángel Gregorio. Topaz: An open-source interconnection network simulator for chip multiprocessors and supercomputers. In *Networks on Chip (NoCS), 2012 Sixth IEEE/ACM International Symposium on*, pages 99–106. IEEE, 2012. DOI: 10.1109/nocs.2012.19.

[2] Pablo Abad, Valentin Puente, and José Ángel Gregorio. MRR: Enabling fully adaptive multicast routing for CMP interconnection networks. In *International Symposium on High Performance Computer Architecture*, pages 355–366, 2009. DOI: 10.1109/hpca.2009.4798273.

[3] Pablo Abad, Valentin Puente, José Angel Gregorio, and Pablo Prieto. Rotary router: An efficient architecture for CMP interconnection networks. In *Proc. of the International Symposium on Computer Architecture*, pages 116–125, June 2007. DOI: 10.1145/1250662.1250678.

[4] Sergi Abadal, Albert Cabellos-Aparicio, Eduard Alarcón, and Josep Torrellas. WiSync: An architecture for fast synchronization through on-chip wireless communication. In *Proc. of the Twenty-First International Conference on Architectural Support for Programming Languages and Operating Systems, ASPLOS'16, Atlanta, GA, USA, April 2-6, 2016*, pages 3–17, 2016. DOI: 10.1145/2872362.2872396.

[5] Sergi Abadal, Benny Sheinman, Oded Katz, Ofer Markish, Danny Elad, Yvan Fournier, Damian Roca, Mauricio Hanzich, Guillaume Houzeaux, Mario Nemirovsky, et al. Broadcast-enabled massive multicore architectures: A wireless rf approach. *IEEE Micro*, 35(5):52–61, 2015. DOI: 10.1109/mm.2015.123.

[6] Mohamed Abdelfattah and Vaughn Betz. Design tradeoffs for hard and soft fpga-based networks-on-chip. In *International Conference on Field-Programmable Technology*, 2012. DOI: 10.1109/fpt.2012.6412118.

[7] Mohamed Abdelfattah and Vaughn Betz. The power of communication: Energy-efficient NoCs for FPGAs. In *International Conference on Field-Programmable Logic and Applications*, 2013. DOI: 10.1109/fpl.2013.6645496.

[8] Mohamed Abdelfattah and Vaughn Betz. Power analysis of embedded NoCs on FPGAs and comparison to custom buses. *IEEE Trans. on VLSI*, January 2016. DOI: 10.1109/tvlsi.2015.2397005.

[9] Mohamed Abdelfattah, Andrew Bitar, and Vaughn Betz. Take the highway: Design for embedded NoCs on FPGAs. In *International Symposium on Field Programmable Gate Arrays*, 2015. DOI: 10.1145/2684746.2689074.

[10] Nilmini Abeyratne, Reetuparna Das, Qingkun Li, Korey Sewell, Bharan Giridhar, Ronald G. Dreslinski, David Blaauw, and Trevor Mudge. Scaling towards kilo-core processors with asymmetric high radix topologies. In *Proc. of the International Symposium on High Performance Computer Architecture*, 2013. DOI: 10.1109/hpca.2013.6522344.

[11] Ahmed K Abousamra, Rami G Melhem, and Alex K Jones. Deja vu switching for multiplane NoCs. In *Networks on Chip (NoCS), 2012 Sixth IEEE/ACM International Symposium on*, pages 11–18. IEEE, 2012. DOI: 10.1109/nocs.2012.9.

[12] Dennis Abts, Abdulla Bataineh, Steve Scott, Greg Faanes, Jim Schwarzmeier, Eric Lundberg, Tim Johnson, Mike Bye, and Gerald Schwoerer. The Cray BlackWidow: a highly scalable vector multiprocessor. In *Proc. of the Conference on Supercomputing*, page 17, 2007. DOI: 10.1145/1362622.1362646.

[13] Dennis Abts, Natalie Enright Jerger, John Kim, Mikko Lipasti, and Dan Gibson. Achieving predictable performance through better memory controller placement in many-core CMPs. In *Proc. of the International Symposium on Computer Architecture*, 2009. DOI: 10.1145/1555754.1555810.

[14] Niket Agarwal, Tushar Krishna, Li-Shiuan Peh, and Niraj K. Jha. GARNET: A detailed on-chip network model inside a full-system simulator. In *Proc. of the IEEE International Symposium on Performance Analysis of Systems and Software*, pages 33–42, April 2009. DOI: 10.1109/ispass.2009.4919636.

[15] Niket Agarwal, Li-Shiuan Peh, and Niraj K. Jha. In-network coherence filtering: Snoopy coherence without broadcasts. In *International Symposium on Microarchitecture*, 2009. DOI: 10.1145/1669112.1669143.

[16] Niket Agarwal, Li-Shiuan Peh, and Niraj K. Jha. In-network snoop ordering (INSO): Snoopy coherence on unordered interconnects. In *Proc. of the International Symposium on High Performance Computer Architecture*, pages 67–78, February 2009. DOI: 10.1109/hpca.2009.4798238.

[17] Junwhan Ahn, Sungpack Hong, Sungjoo Yoo, Onur Mutlu, and Kiyoung Choi. A scalable processing-in-memory accelerator for parallel graph processing. In *Proc. of the 42nd Annual International Symposium on Computer Architecture*, pages 105–117, Portland, OR, June 13–17, 2015. DOI: 10.1145/2749469.2750386.

[18] Minseon Ahn and Eun Jung Kim. Pseudo-circuit: Accelerating communication for on-chip interconnection networks. In *Proc. of the International Symposium on Microarchitecture*, December 2010. DOI: 10.1109/micro.2010.10.

[19] Konstantinos Aisopos, Chia-Hsin Owen Chen, and Li-Shiuan Peh. Enabling system-level modeling of variation-induced faults in networks-on-chips. In *Proc. of the 48th Design Automation Conference*, DAC '11, pages 930–935, New York, NY, 2011. ACM. DOI: 10.1145/2024724.2024931.

[20] Konstantinos Aisopos, Andrew DeOrio, Li-Shiuan Peh, and Valeria Bertacco. ARI-ADNE: agnostic reconfiguration in a disconnected network environment. In *2011 International Conference on Parallel Architectures and Compilation Techniques, PACT*, pages 298–309, Galveston, TX, October 10–14, 2011. DOI: 10.1109/pact.2011.61.

[21] Filipp Akopyan, Jun Sawada, Andrew Cassidy, Rodrigo Alvarez-Icaza, John Arthur, Paul Merolla, Nabil Imam, Yutaka Nakamura, Pallab Datta, Gi-Joon Nam, Brian Taba, Michael Beakes, Bernard Brezzo, Jente B. Kuang, Rajit Manohar, William P. Risk, Bryan Jackson, and Dharmendra S. Modha. TrueNorth: Design and tool flow of a 65 mW 1 million neuron programmable neurosynaptic chip. *IEEE Transactions on Computer-Aided Design of Integrated Circuits and Systems*, 34(10):1537–1557, 2015. DOI: 10.1109/tcad.2015.2474396.

[22] Adrijean Andriahantenaina, Herve Charlery, Alain Greiner, Laurent Mortiez, and Cesar Albenes Zeferino. SPIN: a scalable, packet switched, on-chip micro-network. In *Proc. of the Conference on Design, Automation and Test in Europe*, pages 70–73, 2003. DOI: 10.1109/date.2003.1253808.

[23] Rajeev Balasubramonian Aniruddha N. Udipi, Naveen Muralimanohar. Towards scalable, energy-efficient bus-based on-chip networks. In *Proc. of the International Symposium on High Performance Computer Architecture*, 2010. DOI: 10.1109/hpca.2010.5416639.

[24] Amin Ansari, Asit Mishra, Jianping Xu, and Josep Torrellas. Tangle: Route-oriented dynamic voltage minimization for variation-afflicted, energy-efficient on-chip networks. In *International Symposium on High Performance Computer Architecture*, 2014. DOI: 10.1109/hpca.2014.6835953.

[25] Padma Apparao, Ravi Iyer, Xiaomin Zhang, Don Newell, and Tom Adelmeyer. Characterization and analysis of a server consolidation benchmark. In *Proc. of the International Conference on Virtual Execution Environments*, pages 21–30, 2008. DOI: 10.1145/1346256.1346260.

[26] ARM. AMBA open specifications. https://www.arm.com/products/amba-open-specifications.php.

[27] ARM. AMBA specifications. https://www.arm.com/products/system-ip/amba-sp ecifications.

[28] ARM. Core link interconnect. https://www.arm.com/products/system-ip/corelin k-interconnect.

[29] Arteris. Flexnoc. http://www.arteris.com/flexnoc.

[30] Krste Asanovic, Rastislav Bodik, James Demmel, Tony Keaveny, Kurt Keutzer, John Kubiatowicz, Nelson Morgan, David Patterson, Koushik Sen, John Wawrzynek, David Wessel, and Katherine Yelick. A view of the parallel computing landscape. *Communications of the ACM*, 52(10):56–67, 2009. DOI: 10.1145/1562764.1562783.

[31] Infiniband Trade Association. http://www.infinibandta.org/about.

[32] Todd Austin, Valeria Bertacco, Scott Mahlke, and Yu Cao. Reliable systems on unreliable fabrics. *IEEE Design and Test of Computers*, 25(4):322–332, July 2008. DOI: 10.1109/mdt.2008.107.

[33] Jonathan Bachrach, Huy Vo, Brian Richards, Yunsup Lee, Andrew Waterman, Rimas Avižienis, John Wawrzynek, and Krste Asanović. Chisel: constructing hardware in a scala embedded language. In *Proc. of the 49th Annual Design Automation Conference*, pages 1216–1225. ACM, 2012. DOI: 10.1145/2228360.2228584.

[34] Mario Badr and Natalie Enright Jerger. SynFull: Synthetic traffic models capturing cache coherent behaviour. In *Proc. of the International Symposium on Computer Architecture*, 2014. DOI: 10.1109/isca.2014.6853236.

[35] Ali Bakhoda, John Kim, and Tor M. Aamodt. Throughput-effective on-chip networks for manycore accelerators. In *Proc. of the International Symposium on Microarchitecture*, 2010. DOI: 10.1109/micro.2010.50.

[36] Ali Bakhoda, George L Yuan, Wilson WL Fung, Henry Wong, and Tor M Aamodt. Analyzing CUDA workloads using a detailed GPU simulator. In *Performance Analysis of Systems and Software, 2009. ISPASS 2009. IEEE International Symposium on*, pages 163–174. IEEE, 2009. DOI: 10.1109/ispass.2009.4919648.

[37] Rajeev Balasubramonian, Naveen Muralimanohar, Karthik Ramani, Liqun Cheng, and John Carter. Leveraging wire properties at the microarchitecture level. *IEEE Micro*, 26(6):40–52, Nov/Dec 2006. DOI: 10.1109/mm.2006.123.

[38] James Balfour and William J. Dally. Design tradeoffs for tiled CMP on-chip networks. In *Proc. of the International Conference on Supercomputing*, pages 187–198, 2006. DOI: 10.1145/2591635.2667187.

[39] Jonathan Balkind, Michael McKeown, Yaosheng Fu, Tri Nguyen, Yanqi Zhou, Alexey Lavrov, Mohammad Shahrad, Adi Fuchs, Samuel Payne, Xiaohua Liang, Matthew Matl, and David Wentzlaff. OpenPiton: An open source manycore research framework. In *Proc. of the Twenty-First International Conference on Architectural Support for Programming Languages and Operating Systems*, pages 217–232. ACM, 2016. DOI: 10.1145/2872362.2872414.

[40] Luiz A. Barroso, Kourosh Gharachorloo, Robert McNamara, Andreas Nowatzyk, Shaz Qadeer, Barton Sano, Scott Smith, Robert Stets, and Ben Verghese. Piranha: a scalable architecture based on single-chip multiprocessing. In *Proc. of the International Symposium on Computer Architecture*, pages 282–293, 2000. DOI: 10.1109/isca.2000.854398.

[41] Edith Beigné, Fabien Clermidy, Pascal Vivet, Alain Clouard, and Marc Renaudin. An asynchronous noc architecture providing low latency service and its multi-level design framework. In *Asynchronous Circuits and Systems, 2005. ASYNC 2005. Proceedings. 11th IEEE International Symposium on*, pages 54–63. IEEE, 2005. DOI: 10.1109/async.2005.10.

[42] Luca Benini, Davide Bertozzi, Alessandro Bogliolo, Francesco Menichelli, and Mauro Olivieri. MPARM: Exploring the multi-processor SoC design space with SystemC. *Journal of VLSI Signal Processing Systems*, 41(2):169–182, September 2005. DOI: 10.1007/s11265-005-6648-1.

[43] Luca Benini and Giovanni De Micheli. Powering networks on chips. In *Proc. of the 14th International Symposium on System Synthesis*, pages 33–38, 2001. DOI: 10.1109/isss.2001.957909.

[44] Luca Benini and Giovanni De Micheli. Networks on chips: a new SoC paradigm. *IEEE Computer*, 35(1):70–78, January 2002. DOI: 10.1109/2.976921.

[45] Luca Benini and Giovanni De Micheli. *Networks on Chips: Technology and Tools*. Academic Press, 2006.

[46] Keren Bergman, John Shalf, and Tom Hausken. Optical interconnects and extreme computing. *Optics and Photonics News*, 27(4):32–39, April 2016. DOI: 10.1364/opn.27.4.000032.

[47] Davide Bertozzi, Antoine Jalabert, Srinivasan Murali, Rutuparna Tamhankar, Stergios Stergiou, Luca Benini, and Giovanni De Micheli. NoC synthesis flow for customized domain specific multiprocessor systems-on-chip. *IEEE Transactions on Parallel and Distributed Systems*, 16(2):113–129, February 2005. DOI: 10.1109/tpds.2005.22.

[48] Christian Bienia, Sanjeev Kumar, Jaswinder Pal Singh, and Kai Li. The PARSEC benchmark suite: Characterization and architectural implications. In *Proc. of the 17th International Conference on Parallel Architectures and Compilation Techniques*, pages 72–81, October 2008. DOI: 10.1145/1454115.1454128.

[49] Nathan Binkert, Bradford Beckmann, Gabriel Black, Steven K. Reinhardt, Ali Saidi, Arkaprava Basu, Joel Hestness, Derek R. Hower, Tushar Krishna, Somayeh Sardashti, Rathijit Sen, Korey Sewell, Muhammad Shoaib, Nilay Vaish, Mark D. Hill, and David A. Wood. The gem5 simulator. *SIGARCH Computer Architecture News*, 39(2):1–7, 2011. DOI: 10.1145/2024716.2024718.

[50] Tobias Bjerregaard and Shankar Mahadevan. A survey of research and practices of network-on-chip. *ACM Computer Surveys*, 38(1), 2006. DOI: 10.1145/1132952.1132953.

[51] Paul Bogdan, Radu Marculescu, and Siddharth Jain. Dynamic power management for multidomain system-on-chip platforms: an optimal control approach. *ACM Transactions on Design Automation of Electronic Systems (TODAES)*, 18(4):46, 2013. DOI: 10.1145/2504904.

[52] Haseeb Bokhari, Haris Javaid, Muhammad Shafique, Jörg Henkel, and Sri Parameswaran. Malleable NoC: dark silicon inspired adaptable network-on-chip. In *Proc. of the 2015 Design, Automation & Test in Europe Conference & Exhibition*, pages 1245–1248. EDA Consortium, 2015. DOI: 10.7873/date.2015.0694.

[53] Evgeny Bolotin, Israel Cidon, Ran Ginosar, and Avinoam Kolodny. QNoC: QoS architecture and design processor for cost-effective network on chip. *Special issue on Networks on Chip, The Journal of Systems Architecture*, 50(2):105–128, February 2004. DOI: 10.1016/j.sysarc.2003.07.004.

[54] Evgeny Bolotin, Israel Cidon, Ran Ginosar, and Avinoam Kolodony. Routing table minimization for irregular mesh NoCs. In *Proc. of the Conference on Design, Automation and Test in Europe*, pages 942–947, 2007. DOI: 10.1109/date.2007.364414.

[55] Anja Boos, Luca Ramini, Ulf Schlichtmann, and Davide Bertozzi. Proton: An automatic place-and-route tool for optical networks-on-chip. In *International Conference on Computer-Aided Design*, pages 138–145. IEEE, 2013. DOI: 10.1109/iccad.2013.6691109.

[56] Aaron Carpenter, Jianyun Hu, Ovunc Kocabas, Michael Huang, and Hui Wu. Enhancing effective throughput for transmission line-based bus. In *Proc. of the 39th Annual International Symposium on Computer Architecture*, ISCA '12, pages 165–176, 2012. DOI: 10.1109/isca.2012.6237015.

[57] Aaron Carpenter, Jianyun Hu, Jie Xu, Michael Huang, and Hui Wu. A case for globally shared-medium on-chip interconnect. In *Proc. of the 38th Annual International Symposium on Computer Architecture*, ISCA '11, pages 271–282, 2011. DOI: 10.1145/2000064.2000097.

[58] Mario R Casu and Paolo Giaccone. Rate-based vs delay-based control for DVFS in NoC. In *2015 Design, Automation & Test in Europe Conference & Exhibition (DATE)*, pages 1096–1101. IEEE, 2015. DOI: 10.7873/date.2015.0613.

[59] Jeremy Chan and Sri Parameswaran. NoCGEN: A template based reuse methodology for networks on chip architecture. In *VLSI Design, 2004. Proceedings. 17th International Conference on*, pages 717–720. IEEE, 2004. DOI: 10.1109/icvd.2004.1261011.

[60] M. Frank Chang, Jason Cong, Adam Kaplan, Chunyue Liu, Mishali Naik, Jagannath Premkumar, Glenn Reinman, Eran Socher, and Sai-Wang Tam. Power reduction of CMP communication networks via RF-interconnects. In *Proc. of the 41st Annual International Symposium on Microarchitecture*, pages 376–387, November 2008. DOI: 10.1109/micro.2008.4771806.

[61] M. Frank Chang, Jason Cong, Adam Kaplan, Mishali Naik, Glenn Reinman, Eran Socher, and Sai-Wang Tam. CMP network-on-chip overlaid with multi-band RF-interconnect. In *Proc. of the 14th International Symposium on High-Performance Computer Architecture*, pages 191–202, February 2008. DOI: 10.1109/hpca.2008.4658639.

[62] Yuan-Ying Chang, Yoshi Shih-Chieh Huang, Matthew Poremba, Vijaykrishnan Narayanan, Yuan Xie, and Chung-Ta King. TS-Router: On maximizing the quality-of-allocation in the on-chip network. In *Proc. of the International Symposium on High Performance Computer Architecture*, 2013. DOI: 10.1109/hpca.2013.6522335.

[63] Shuai Che, Michael Boyer, Jiayuan Meng, David Tarjan, Jeremy W. Sheaffer, Sang-Ha Lee, and Kevin Skadron. Rodinia: A benchmark suite for heterogeneous computing. In *Proc. of the 2009 IEEE International Symposium on Workload Characterization (IISWC)*, IISWC '09, pages 44–54, Washington, DC, USA, 2009. IEEE Computer Society. DOI: 10.1109/iiswc.2009.5306797.

[64] Chia-Hsin Chen. *Design and Implementation of Low-latency, Low-power Reconfigurable On-Chip Networks*. Ph.D. thesis, Massachusetts Institute of Technology, 2016.

[65] Chia-Hsin Owen Chen, Sunghyun Park, Tushar Krishna, and Li-Shiuan Peh. A low-swing crossbar and link generator for low-power networks-on-chip. In *International Conference on Computer-Aided Design*, 2011. DOI: 10.1109/iccad.2011.6105418.

[66] L. Chen, D. Zhu, M. Pedram, and T. M. Pinkston. Simulation of NoC power-gating: Requirements, optimizations,and the Agate simulator. *Journal of Parallel and Distributed Computing*, 2016. DOI: 10.1016/j.jpdc.2016.03.006.

[67] Lizhong Chen and Timothy M Pinkston. Nord: Node-router decoupling for effective power-gating of on-chip routers. In *Proc. of the 2012 45th Annual IEEE/ACM International Symposium on Microarchitecture*, pages 270–281. IEEE Computer Society, 2012. DOI: 10.1109/micro.2012.33.

[68] Lizhong Chen and Timothy M. Pinkston. Worm-bubble flow control. In *Proc. of the International Symposium on High Performance Computer Architecture*, February 2013. DOI: 10.1109/hpca.2013.6522333.

[69] Lizhong Chen, Lihang Zhao, Ruisheng Wang, and Timothy Mark Pinkston. MP3: Minimizing performance penalty for power-gating of clos network-on-chip. In *International Symposium on High Performance Computer Architecture*, 2014. DOI: 10.1109/hpca.2014.6835940.

[70] Lizhong Chen, Di Zhu, Massoud Pedram, and Timothy M Pinkston. Power punch: Towards non-blocking power-gating of NoC routers. In *2015 IEEE 21st International Symposium on High Performance Computer Architecture (HPCA)*, pages 378–389. IEEE, 2015. DOI: 10.1109/hpca.2015.7056048.

[71] Xi Chen, Zheng Xu, Hyungjun Kim, Paul V Gratz, Jiang Hu, Michael Kishinevsky, Umit Ogras, and Raid Ayoub. Dynamic voltage and frequency scaling for shared resources in multicore processor designs. In *Proc. of the 50th Annual Design Automation Conference*, page 114. ACM, 2013. DOI: 10.1145/2463209.2488874.

[72] Yu-Hsin Chen, Tushar Krishna, Joel Emer, and Vivienne Sze. Eyeriss: An Energy-Efficient Reconfigurable Accelerator for Deep Convolutional Neural Networks. In *IEEE International Solid-State Circuits Conference, ISSCC 2016, Digest of Technical Papers*, pages 262–263, 2016. DOI: 10.1109/jssc.2016.2616357.

[73] Andrew A. Chien and Jae H. Kim. Planar-adaptive routing: low-cost adaptive networks for multiprocessors. In *Proc. of the International Symposium on Computer Architecture*, pages 268–277, 1992. DOI: 10.1109/isca.1992.753323.

[74] Ge-Ming Chiu. The odd-even turn model for adaptive routing. *IEEE Transactions on Parallel and Distributed Systems*, pages 729–738, July 2000. DOI: 10.1109/71.877831.

[75] Myong Hyon Cho, Mieszko Lis, Keun Sup Shim, Michel Kinsy, Tina Wen, and Srinivas Devadas. Oblivious routing on on-chip bandwidth-adaptive networks. In *Proc. of the International Conference on Parallel Architecture and Compilation Techniques*, 2009. DOI: 10.1109/pact.2009.41.

[76] Eric S. Chung, James C. Hoe, and Ken Mai. CoRAM: An in-fabric memory architecture for fpga-based computing. In *International Symposium on Field-Programmable Gate Arrays*, 2011. DOI: 10.1145/1950413.1950435.

[77] Mark J. Cianchetti, Joseph C. Kerekes, and David H. Albonesi. Phastlane: A rapid transit optical routing network. In *International Symposium on Computer Architecture*, 2009. DOI: 10.1145/1555754.1555809.

[78] Christopher Condrat, Priyank Kalla, and Steve Blair. Crossing-aware channel routing for integrated optics. *TCAD*, 33(6):814–825, 2014. DOI: 10.1109/tcad.2014.2317575.

[79] Christopher Condrat, Priyank Kalla, and Steve Blair. Thermal-aware synthesis of integrated photonic ring resonators. In *International Conference on Computer-Aided Design*, 2014. DOI: 10.1109/iccad.2014.7001405.

[80] Kypros Constantinides, Stephen Plaza, Jason Blome, Bin Zhang, Valeria Bertacco, Scott Mahlke, Todd Austin, and Michael Orshansky. BulletProof: A defect tolerant CMP switch architecture. In *Proc. of the International Symposium on High Performance Computer Architecture*, pages 5–16, 2006. DOI: 10.1109/hpca.2006.1598108.

[81] Pat Conway and Bill Hughes. The AMD Opteron Northbridge architecture, present and future. *IEEE Micro Magazine*, 27:10–21, March 2007. DOI: 10.1109/MM.2007.43.

[82] Marcello Coppola. Spidergon STNoC: The technology that adds value to your system. In *Hot Chips 22 Symposium (HCS), 2010 IEEE*, pages 1–39. IEEE, 2010. DOI: 10.1109/hotchips.2010.7480082.

[83] Marcello Coppola, Riccardo Locatelli, Giuseppe Maruccio, Lorenzo Pieralisi, and A. Scandurra. Spidergon: a novel on chip communication network. In *International Symposium on System on Chip*, page 15, November 2004. DOI: 10.1109/issoc.2004.1411133.

[84] Cisco CRS-1. http://www.cisco.com.

[85] D. E. Culler and J. P. Singh. *Parallel Computer Architecture: A Hardware/Software Approach*. Morgan Kaufmann Publishers Inc., 1999.

[86] William Dally and Brian Towles. *Principles and Practices of Interconnection Networks*. Morgan Kaufmann Pub., San Francisco, CA, 2003.

[87] William J. Dally. Virtual-channel flow control. In *Proc. of the International Symposium on Computer Architecture*, 1990. DOI: 10.1109/isca.1990.134508.

[88] William J. Dally. Express cubes: Improving the performance of k-ary n-cube interconnection networks. *IEEE Transactions on Computers*, 40(9):1016–1023, September 1991. DOI: 10.1109/12.83652.

[89] William J. Dally and Hiromichi Aoki. Deadlock-free adaptive routing in multicomputer networks using virtual channels. *IEEE Transactions on Parallel and Distributed Systems*, 4(4):466–475, 1993. DOI: 10.1109/71.219761.

[90] William J. Dally, Larry R. Dennison, David Harris, Kinhong Kan, and Thucydides Xanthopoulos. The reliable router: A reliable and high-performance communication substrate for parallel computers. In *Proc. of the First International Workshop on Parallel Computer Routing and Communication*, pages 241–255, 1994. DOI: 10.1007/3-540-58429-3_41.

[91] William J. Dally, J. A. Stuart Fiske, John S. Keen, Richard A. Lethin, Michael D. Noakes, Peter R. Nuth, Roy E. Davison, and Gregory A. Fyler. The message-driven processor – a multicomputer processing node with efficient mechanisms. *IEEE Micro*, 12(2):23–39, April 1992. DOI: 10.1109/40.127581.

[92] William J. Dally and John W. Poulton. *Digital Systems Engineering*. Cambridge University Press, 1998. DOI: 10.1017/cbo9781139166980.

[93] William J. Dally and Charles L. Seitz. The torus routing chip. *Journal of Distributed Computing*, 1(3):187–196, 1986. DOI: 10.1007/bf01660031.

[94] William J. Dally and Charles L. Seitz. Deadlock-free message routing in multiprocessor interconnection networks. *IEEE Transactions on Computers*, 36(5):547–553, 1987. DOI: 10.1109/tc.1987.1676939.

[95] William J. Dally and Brian Towles. Route packets, not wires: On-chip interconnection networks. In *Proc. of the 38th Conference on Design Automation*, pages 684–689, 2001. DOI: 10.1109/dac.2001.935594.

[96] Reetuparna Das, Rachata Ausavarungnirun, Onur Mutlu, Akhilesh Kumar, and Mani Azimi. Application-to-core mapping policies to reduce memory system interference in multi-core systems. In *International Sympoisum on High Performance Computer Architecture*, pages 107–118, 2013. DOI: 10.1109/hpca.2013.6522311.

[97] Reetuparna Das, Soumya Eachempati, Asit K. Mishra, N. Vijaykrishnan, and Chita R. Das. Design and evaluation of hierarchical on-chip network topologies for next generation CMPs. In *Proc. of the International Symposium on High Performance Computer Architecture*, pages 175–186, February 2009.

[98] Reetuparna Das, Onur Mutlu, Thomas Moscibroda, and Chita Das. Application-aware priorization mechanisms for on-chip networks. In *Proc. of the International Symposium on Microarchitecture*, 2009. DOI: 10.1145/1669112.1669150.

[99] Reetuparna Das, Onur Mutlu, Thomas Moscibroda, and Chita Das. Aergia: Exploiting packet latency slack in on-chip networks. In *Proc. of the International Symposium on Computer Architecture*, 2010. DOI: 10.1145/1815961.1815976.

[100] Reetuparna Das, Satish Narayanasamy, Sudhir Satpathy, and Ronald Dreslinski. Catnap: Energy proportional multiple network-on-chip. In *Proc. of the International Symposium on Computer Architecture*, 2013. DOI: 10.1145/2508148.2485950.

[101] Bhavya K. Daya, Chia-Hsin Owen Chen, Suvinay Subramanian, Woo-Cheol Kwon, Sunghyun Park, Tushar Krishna, Jim Holt, Anantha P. Chandrakasan, and Li-Shiuan Peh. SCORPIO: A 36-core research chip demonstrating snoopy coherence on a scalable mesh NoC with in-network ordering. In *ACM/IEEE 41st International Symposium on Computer Architecture, ISCA*, pages 25–36, Minneapolis, MN, June 14–18, 2014. DOI: 10.1109/isca.2014.6853232.

[102] Martin De Prycker. *Asynchronous Transfer Mode: Solution for Broadband ISDN*, 3rd ed., Prentice Hall, 1995.

[103] Sujay Deb, Amlan Ganguly, Partha Pratim Pande, Benjamin Belzer, and Deukhyoun Heo. Wireless noc as interconnection backbone for multicore chips: Promises and challenges. *IEEE Journal on Emerging and Selected Topics in Circuits and Systems*, 2(2):228–239, 2012. DOI: 10.1109/jetcas.2012.2193835.

[104] Duo Ding, Yilin Zhang, Haiyu Huang, Ray T. Chen, and David Z. Pan. O-router: an optical routing framework for low power on-chip silicon nano-photonic integration. In *Proc. of the Design Automation Conference*. ACM, 2009. DOI: 10.1145/1629911.1629983.

[105] Dominic DiTomaso, Avinash Kodi, and Ahmed Louri. QORE: A fault tolerant network-on-chip architecture with power-efficient quad-function channel (QFC) buffers. In *International Symposium on High Performance Computer Architecture*, 2014. DOI: 10.1109/hpca.2014.6835942.

[106] T. Dorta, J. Jimnez, J. L. Martn, U. Bidarte, and A. Astarloa. Overview of fpga-based multiprocessor systems. In *International Conference on Reconfigurable Computing and FPGAs*, 2009. DOI: 10.1109/reconfig.2009.15.

[107] Zidong Du, Robert Fasthuber, Tianshi Chen, Paolo Ienne, Ling Li, Tao Luo, Xiaobing Feng, Yunji Chen, and Olivier Temam. ShiDianNao: Shifting vision processing closer to the sensor. In *International Symposium on Computer Architecture*, 2015. DOI: 10.1145/2749469.2750389.

[108] José Duato. A new theory of deadlock-free adaptive routing in wormhole networks. *IEEE Transactions on Parallel and Distributed Systems*, 4(12):1320–1331, December 1993. DOI: 10.1109/spdp.1993.395549.

[109] José Duato. A necessary and sufficient condition for deadlock-free adaptive routing in wormhole networks. *IEEE Transactions on Parallel and Distributed Systems*, 6(10):1055–1067, Oct 1995. DOI: 10.1109/71.473515.

[110] José Duato, Sudhakar Yalamanchili, and Lionel M. Ni. *Interconnection Networks: An Engineering Approach*, 2nd ed., Morgan Kaufmann, 2003.

[111] Noel Eisley, Li-Shiuan Peh, and Li Shang. In-network cache coherence. In *Proc. of the 39th International Symposium on Microarchitecture*, pages 321–332, December 2006. DOI: 10.1109/micro.2006.27.

[112] Natalie Enright Jerger, Ajaykumar Kannan, Zimo Li, and Gabriel H. Loh. NoC architectures for silicon interposer systems. In *International Symposium on Microarchitecture*, 2014. DOI: 10.1109/micro.2014.61.

[113] Natalie Enright Jerger, Li-Shiuan Peh, and Mikko H. Lipasti. Circuit-switched coherence. In *Proc. of the International Network on Chip Symposium*, pages 193–202, April 2008. DOI: 10.1109/nocs.2008.4492738.

[114] Natalie Enright Jerger, Li-Shiuan Peh, and Mikko H. Lipasti. Virtual circuit tree multicasting: A case for on-chip hardware multicast support. In *International Symposium on Computer Architecture*, pages 229–240, June 2008. DOI: 10.1109/isca.2008.12.

[115] Natalie Enright Jerger, Li-Shiuan Peh, and Mikko H. Lipasti. Virtual tree coherence: Leveraging regions and in-network multicast trees for scalable cache coherence. In *Proc. of the 41st International Symposium on Microarchitecture*, pages 35–46, November 2008. DOI: 10.1109/micro.2008.4771777.

[116] Natalie Enright Jerger, Dana Vantrease, and Mikko H. Lipasti. An evaluation of server consolidation workloads for multi-core designs. In *IEEE International Symposium on Workload Consolidation*, pages 47–56, September 2007. DOI: 10.1109/iiswc.2007.4362180.

[117] Chris Fallin, Chris Craik, and Onur Mutlu. CHIPPER: A low-complexity bufferless deflection router. In *Proc. of the International Symposium on High Performance Computer Architecture*, 2011. DOI: 10.1109/hpca.2011.5749724.

[118] Farzad Fatollahi-Fard, David Donofrio, George Michelogiannakis, and John Shalf. OpenSoC fabric: On-chip network generator. In *IEEE International Symposium on Performance Analysis of Systems and Software, ISPASS*, pages 194–203, 2016. DOI: 10.1109/ispass.2016.7482094.

[119] David Fick, Andrew DeOrio, Gregory Chen, Valeria Bertacco, Dennis Sylvester, and David Blaauw. A highly resilient routing algorithm for fault-tolerant NoCs. In *Proc. of the Conference on Design, Automation and Test in Europe*, DATE '09, pages 21–26, 3001 Leuven, Belgium, Belgium, 2009. European Design and Automation Association. DOI: 10.1109/date.2009.5090627.

[120] David Fick, Andrew DeOrio, Jin Hu, Valeria Bertacco, David Blaauw, and Dennis Sylvester. Vicis: A reliable network for unreliable silicon. In *Proc. of the 46th Annual Design Automation Conference, DAC'09*, pages 812–817, New York, NY, 2009. ACM. DOI: 10.1145/1629911.1630119.

[121] Finisar. Optimized fiber optics solutions for data center applications. http://www.finisar.com/markets/data-center.

[122] J. Flich, A. Mejia, P. Lopez, and J. Duato. Region-based routing: An efficient routing mechanism to tackle unreliable hardware in network on chips. In *Proc. of the First International Symposium on Networks-on-Chip, NOCS'07*, pages 183–194, Washington, DC, 2007. IEEE Computer Society. DOI: 10.1109/nocs.2007.39.

[123] Jose Flich, Andres Mejia, Pedro López, and José Duato. Region-based routing: An efficient routing mechanism to tackle unreliable hardware in networks on chip. In *Proc. of the Network on Chip Symposium*, pages 183–194, May 2007. DOI: 10.1109/nocs.2007.39.

[124] Jose Flich, Samuel Rodrigo, and José Duato. An efficient implementation of distributed routing algorithms for NoCs. In *Proc. of the International Network On Chip Symposium*, pages 87–96, April 2008. DOI: 10.1109/nocs.2008.4492728.

[125] Binzhang Fu, Yinhe Han, Jun Ma, Huawei Li, and Xiaowei Li. An abacus turn model for time/space-efficient reconfigurable routing. In *Proc. of the International Symposium on Computer Architecture*, June 2011. DOI: 10.1145/2000064.2000096.

[126] Mike Galles. Scalable pipelined interconnect for distributed endpoint routing: The SGI SPIDER chip. In *Proc. of Hot Interconnects Symposium IV*, pages 141–146, 1996.

[127] Alan Gara, Matthias A. Blumrich, Dong Chen, George L.-T. Chiu, Paul Coteus, Mark E. Giampapa, Ruud A. Haring, Philip Heidelberger, Dirk Hoenicke, Gerard V. Kopcsay, Thomas A. Liebsch, Martin Ohmacht, Burkhard D. Steinmacher-Burow, Todd Takken, and Pavlos Vranas. Overview of the Blue Gene/L system architecture. *IBM Journal of Research and Developement*, 49(2–3):195–212, 2005. DOI: 10.1147/rd.492.0195.

[128] Patrick T. Gaughan and Sudhakar Yalamanchili. Pipelined circuit-switching: a fault-tolerant variant of wormhole routing. In *Proc. of the Symposium on Parallel and Distributed Processing*, pages 148–155, December 1992. DOI: 10.1109/spdp.1992.242751.

[129] N. Genko, D. Atienza, G. De Micheli, J. Mendias, R. Hermida, and F. Catthoor. A complete network-on-chip emulation framework. In *Proc. of the Conference on Design Automation and Test in Europe*, pages 246–251, March 2005. DOI: 10.1109/date.2005.5.

[130] R. Gindin, I. Cidon, and I. Keidar. NoC-based FPGA: Architecture and routing. In *International Symposium on Networks-on-Chip*, 2007. DOI: 10.1109/nocs.2007.31.

[131] Christopher J. Glass and Lionel M. Ni. The turn model for adaptive routing. In *Proc. of the International Symposium on Computer Architecture*, pages 278–287, May 1992. DOI: 10.1109/isca.1992.753324.

[132] Nitin Godiwala, Jud Leonard, and Matthew Reilly. A network fabric for scalable multiprocessor systems. In *Proc. of the Symposium on Hot Interconnects*, pages 137–144, 2008. DOI: 10.1109/hoti.2008.24.

[133] Kees Goossens, Martijn Bennebroek, Jae Young Hur, and Muhammad Aqeel Wahlah. Hardwired networks on chip in FPGAs to unify functional and configuration interconnects. In *International Symposium on Networks-on-Chip*, 2008. DOI: 10.1109/nocs.2008.4492724.

[134] Kees Goossens, John Dielissen, Om Prakash Gangwal, Santiago Gonzalez Pestana, Andrei Radulescu, and Edwin Rijpkema. A design flow for application-specific networks on chip with guaranteed performance to accelerate SoC design and verification. In *Proc. of the Design, Automation and Test in Europe Conference*, pages 1182–1187, March 2005. DOI: 10.1109/date.2005.11.

[135] Kees Goossens, John Dielissen, and Andrei Radulescu. Æthereal network on chip: Concepts, architectures, and implementations. *IEEE Design and Test*, 22(5):414–421, September 2005. DOI: 10.1109/mdt.2005.99.

[136] N. Goswami, Z. Li, R. Shankar, and T. Li. Exploring silicon nanophotonics in throughput architecture. *IEEE Design & Test*, 31(5):18–27, 2014. DOI: 10.1109/mdat.2014.2348312.

[137] Paul Gratz, Boris Grot, and Stephen W. Keckler. Regional congestion awareness for load balance in networks-on-chip. In *Proc. of the 14th IEEE International Symposium on High Performance Computer Architecture*, pages 203–214, February 2008. DOI: 10.1109/hpca.2008.4658640.

[138] Paul Gratz, Changkyu Kim, Robert G. McDonald, Stephen W. Keckler, and Doug Burger. Implementation and evaluation of on-chip network architectures. In *IEEE International Conference on Computer Design*, pages 477–484, October 2006. DOI: 10.1109/iccd.2006.4380859.

[139] Paul Gratz, Karthikeyan Sankaralingam, Heather Hanson, Premkishore Shivakumar, Robert G. McDonald, Stephen W. Keckler, and Doug Burger. Implementation and evaluation of a dynamically routed processor operand network. In *Proc. of the International Network on Chip Symposium*, pages 7–17, 2007. DOI: 10.1109/nocs.2007.23.

[140] C. Grecu, A. Ivanov, R. Saleh, and P. P. Pande. Testing network-on-chip communication fabrics. *Transactions on Computer-Aided Design of Integrated Circuits and Systems*, 26(12):2201–2214, December 2007. DOI: 10.1109/tcad.2007.907263.

[141] Boris Grot, Joel Hestness, Stephen W. Keckler, and Onur Mutlu. Express cube topologies for on-chip networks. In *Proc. of the International Symposium on High Performance Computer Architecture*, pages 163–174, February 2009. DOI: 10.1109/hpca.2009.4798251.

[142] Boris Grot, Joel Hestness, Stephen W. Keckler, and Onur Mutlu. Kilo-NOC: a heterogeneous network-on-chip for scalabilty and service guarantees. In *Proc. of the International Symposium on Computer Architecture*, 2011. DOI: 10.1145/2000064.2000112.

[143] Boris Grot, Stephen W. Keckler, and Onur Mutlu. Preemptive virtual clock: A flexible, efficient and cost-effective QoS scheme for networks-on-chip. In *Proc. of the International Symposium on Microarchitecture*, 2009. DOI: 10.1145/1669112.1669149.

[144] SMART Interconnects Group. Flexsim 1.2. http://ceng.usc.edu/smart/tools.htm.

[145] Junli Gu, Steven Lumetta, Rakesh Kumar, and Yihe Sun. MOPED: Orchestrating interprocess message data on CMPs. In *International Symposium on High Performance Computer Architecture*, 2011. DOI: 10.1109/hpca.2011.5749721.

[146] S. Hassan and S. Yalamanchili. Centralized buffer router: A low latency, low power router for high radix NoCs. In *Proc. of the International Symposium on Networks on Chip*, 2013. DOI: 10.1109/nocs.2013.6558397.

[147] Syed Minhaj Hassan and Sudhakar Yalamanchili. Bubble sharing: Area and energy efficient adaptive routers using centralized buffers. In *2014 International Symposium on Networks-on-Chip (NOCS)*, September 2014. DOI: 10.1109/nocs.2014.7008770.

[148] Mitchell Hayenga, Natalie Enright Jerger, and Mikko Lipasti. SCARAB: A single cycle adaptive routing and bufferless network. In *Proc. of the International Symposium on Microarchitecture*, December 2009. DOI: 10.1145/1669112.1669144.

[149] Mitchell Hayenga and Mikko Lipasti. The NoX router. In *Proc. of the International Symposium on Microarchitecture*, 2011. DOI: 10.1145/2155620.2155626.

[150] Gilbert Hendry, Johnnie Chan, Luca P Carloni, and Keren Bergman. VANDAL: A tool for the design specification of nanophotonic networks. In *DATE*, pages 1–6. IEEE, 2011. DOI: 10.1109/date.2011.5763133.

[151] Jörg Henkel, Wayne Wolf, and Srimat T. Chakradhar. On-chip networks: A scalable, communication-centric embedded system design paradigm. In *Proc. of VLSI Design*, pages 845–851, January 2004. DOI: 10.1109/icvd.2004.1261037.

[152] John L. Hennessy and David A. Patterson. *Computer Architecture: A Quantitative Approach*, 4th ed., Morgan Kaufmann Publishers Inc., 2006.

[153] Robert Hesse and Natalie Enright Jerger. Improving DVFS in NoCs with coherence prediction. In *Proc. of the 9th International Symposium on Networks-on-Chip*, page 24. ACM, 2015. DOI: 10.1145/2786572.2786595.

[154] Robert Hesse, Jeff Nicholls, and Natalie Enright Jerger. Fine-grained bandwidth adaptivity in networks-on-chip using bidirectional channels. In *Proc. of the 6th International Symposium on Networks-on-Chip*, 2012. DOI: 10.1109/nocs.2012.23.

[155] Joel Hestness, Boris Grot, and Stephen W. Keckler. Netrace: dependency-driven trace-based network-on-chip simulation. In *Third International Workshop on Network on Chip Architectures, NoCArc'10*, pages 31–36, 2010. DOI: 10.1145/1921249.1921258.

[156] Wai Hong Ho and Timothy Mark Pinkston. A methodology for designing efficient on-chip interconnects on well-behaved communication patterns. In *Proc. of the International Symposium on High Performance Computer Architecture*, pages 377–388, February 2003. DOI: 10.1109/hpca.2003.1183554.

[157] Wai Hong Ho and Timothy Mark Pinkston. A design methodology for efficient application-specific on-chip interconnects. *IEEE Transactions on Parallel and Distributed Systems*, 17(2):174–190, February 2006. DOI: 10.1109/tpds.2006.15.

[158] Yatin Hoskote, Sriram Vangal, Arvind Singh, Nitin Borkar, and Shekhar Borkar. A 5-GHz mesh interconnect for a Teraflops processor. *IEEE Micro*, 27(5):51–61, 2007. DOI: 10.1109/mm.2007.4378783.

[159] Jason Howard, Saurabh Dighe, Yatin Hoskote, Sriram Vangal, David Finan, Gregory Ruhl, David Jenkins, Howard Wilson, Nitin Borkar, Gerhard Schrom, Fabrice Pailet, Shailendra Jain, Tiju Jacob, Satish Yada, Sraven Marella, Praveen Salihundam, Vasantha Erraguntla, Michael Konow, Michael Riepen, Guido Droege, Joerg Lindemann, Matthias Gries, Thomas Apel, Kersten Henriss, Tor Lund-Larsen, Sebastian Steibl, Shekhar Borkar, Vivek De, Rob Van Der Wijngaart, and Timothy Mattson. A 48-Core IA-32 Message-Passing Processor with DVFS in 45nm CMOS. In *International Solid-State Circuits Conference*, pages 108–109, 2010. DOI: 10.1109/isscc.2010.5434077.

[160] Jason Howard, Saurabh Dighe, Sriram R. Vangal, Gregory Ruhl, Nitin Borkar, Shailendra Jain, Vasantha Erraguntla, Michael Konow, Michael Riepen, Matthias Gries, Guido Droege, Tor Lund-Larsen, Sebastian Steibl, Shekhar Borkar, Vivek K. De, and Rob Van Der Wijngaart. A 48-core IA-32 processor in 45 nm CMOS using on-die message-passing and DVFS for performance and power scaling. *Journal of Solid-State Circuits*, 46(1):173–183, 2011. DOI: 10.1109/jssc.2010.2079450.

[161] Jingcao Hu and Radu Marculescu. Exploiting the routing flexibility for energy/performance aware mapping of regular NoC architectures. In *Proc. of the Conference on Design, Automation and Test Europe*, pages 688–693, March 2003. DOI: 10.1109/date.2003.1253687.

[162] Jingcao Hu and Radu Marculescu. DyAD–smart routing for networks-on-chip. In *Proc. of the Design Automation Conference*, pages 260–263, June 2004. DOI: 10.1145/996566.996638.

[163] Jingcao Hu and Radu Marculescu. Energy- and performance-aware mapping for regular NoC architectures. *IEEE Transactions on Computer Aided Design for Integrated Circuits Systems*, 24(4):551–562, April 2005. DOI: 10.1109/tcad.2005.844106.

[164] Jingcao Hu, Umit Y. Ogras, and Radu Marculescu. System-level buffer allocation for application specific networks-on-chip router design. *IEEE Transactions on Computer-Aided Design for Integrated Circuits System*, 25(12):2919–2933, December 2006. DOI: 10.1109/tcad.2006.882474.

[165] Paolo Ienne, Patrick Thiran, Giovanni De Micheli, and Frédéric Worm. An adaptive low-power transmission scheme for on-chip networks. In *Proc. of the International Symposium on Systems Synthesis*, pages 92–100, 2002. DOI: 10.1145/581199.581221.

[166] Intel. Intel® QuickPath technology. http://www.intel.com/technology/quickpath.

[167] Intel. From a few cores to many: A Tera-scale computing research overview. http://download.intel.com/research/platform/terascale/terascale_overview/_paper.pdf, 2006.

[168] Syed Ali Raza Jafri, Yu-Ju Hong, Mithuna Thottethodi, and T. N. Vijaykumar. Adaptive flow control for robust performance and energy. In *Proc. of the International Symposium on Microarchitecture*, 2010. DOI: 10.1109/micro.2010.48.

[169] Antoine Jalabert, Srinivasan Murali, Luca Benini, and Giovanni De Micheli. xpipesCompiler: A tool for instantiating application specific networks on chip. In *Proc. of the Conference on Design, Automation and Test in Europe*, volume 2, pages 884–889, February 2004. DOI: 10.1007/978-1-4020-6488-3_12.

[170] Nan Jiang, James Balfour, Daniel U Becker, Brian Towles, William J Dally, George Michelogiannakis, and John Kim. A detailed and flexible cycle-accurate network-on-chip simulator. In *Performance Analysis of Systems and Software (ISPASS), 2013 IEEE International Symposium on*, pages 86–96. IEEE, 2013. DOI: 10.1109/ispass.2013.6557149.

[171] A.P. Jose, G. Patounakis, and K.L. Shepard. Near speed-of-light on-chip interconnects using pulsed current-mode signaling. In *Symposium on VLSI Circuits*, pages 108–111, June 2005. DOI: 10.1109/vlsic.2005.1469345.

[172] Norman P. Jouppi. System implications of integrated photonics. In *Proc. of the International Symposium on Low Power Electronics and Design*, pages 183–184, August 2008. DOI: 10.1145/1393921.1393923.

[173] J. A. Kahl, M. N. Day, H. P. Hofstee, C. R. Johns, T. R. Maeurer, and D. Shippy. Introduction to the Cell multiprocessor. *IBM Journal of Research and Development*, 49(4):589–604, 2005. DOI: 10.1147/rd.494.0589.

[174] Andrew Kahng, Bin Li, Li-Shiuan Peh, and Kambiz Samadi. Orion 2.0: A fast and accurate NoC power and area model for early-stage design space exploration. In *Proc. of the Conference on Design, Automation and Test in Europe*, April 2009. DOI: 10.1109/date.2009.5090700.

[175] Ajaykumar Kannan, Natalie Enright Jerger, and Gabriel H. Loh. Enabling interposer-based disintegration of multi-core processors. In *International Symposium on Microarchitecture*, 2015. DOI: 10.1145/2830772.2830808.

[176] Ajaykumar Kannan, Natalie Enright Jerger, and Gabriel H. Loh. Exploiting interposer technologies to disintegrate and reintegrate multi-core processors for performance and cost. *IEEE Micro Top Picks from Computer Architecture*, 2016. DOI: 10.1109/mm.2016.53.

[177] N. Kapre and J. Gray. Hoplite: Building austere overlay NoCs for FPGAs. In *International Conference on Field-Programmable Logic and Applications*, 2015. DOI: 10.1109/fpl.2015.7293956.

[178] Evangelia Kasapaki, Martin Schoeberl, Rasmus Bo Sørensen, Christoph Müller, Kees Goossens, and Jens Sparsø. Argo: A real-time network-on-chip architecture with an efficient GALS implementation. *IEEE Transactions on Very Large Scale Integration (VLSI) Systems*, 24(2):479–492, 2016. DOI: 10.1109/tvlsi.2015.2405614.

[179] Stefanos Kaxiras and Margaret Martonosi. Computer architecture techniques for power-efficiency. *Synthesis Lectures on Computer Architecture*, 3(1):1–207, 2008. DOI: 10.2200/s00119ed1v01y200805cac004.

[180] Parviz Kermani and Leonar Kleinrock. Virtual cut-through: a new computer communication switching technique. *Computer Networks*, 3(4):267–286, 1979. DOI: 10.1016/0376-5075(79)90032-1.

[181] B. Kim and V. Stojanović. A 4Gb/s/ch 356fj/b 10mm equalized on-chip interconnect with nonlinear charge-injecting transmitter filter and transimpedance receiver in

90nm CMOS technology. In *IEEE Solid-State Circuits Conference*, February 2009. DOI: 10.1109/isscc.2009.4977310.

[182] Changkyu Kim, Doug Burger, and Stephen W. Keckler. An adaptive, non-uniform cache structure for wire-delay dominated on-chip caches. In *Proc. of the 10th International Conference on Architectural Support for Programming Languages and Operating System*, pages 211–222, 2002. DOI: 10.1145/605397.605420.

[183] Dae Hyun Kim, Krit Athikulwongse, Michael Healy, Mohammad Hossain, Moongon Jung, Ilya Khorosh, Gokul Kumar, Young-Joon Lee, Dean Lewis, Tzu-Wei Lin, Chang Liu, Shreepad Panth, Mohit Pathak, Minzhen Ren, Guanhao Shen, Taigon Song, Dong Hyuk Woo, Xin Zhao, Joungho Kim, Ho Choi, Gabriel Loh, Hsien-Hsin Lee, and Sung Kyu Li. 3D-MAPS: 3D massively parallel processor with stacked memory. In *2012 IEEE International Solid-State Circuits Conference*, pages 188–190. IEEE, 2012. DOI: 10.1109/isscc.2012.6176969.

[184] Dae Hyun Kim, Krit Athikulwongse, Michael B Healy, Mohammad M Hossain, Moongon Jung, Ilya Khorosh, Gokul Kumar, Young-Joon Lee, Dean L Lewis, Tzu-Wei Lin, Chang Liu, Shreepad Panth, Mohit Pathak, Minzhen Ren, Guanhao Shen, Taigon Song, Dong Hyuk Woo, Xin Zhao, Joungho Kim, Ho Choi, Gabriel H. Loh, Hsien-Hsin S. Lee, and Sung Kyu Lim. Design and analysis of 3D-MAPS (3D massively parallel processor with stacked memory). *IEEE Transactions on Computers*, 64(1):112–125, 2015. DOI: 10.1109/tc.2013.192.

[185] John Kim. Low-cost router microarchitecture for on-chip networks. In *Proc. of the International Symposium on Microarchitecture*, 2009. DOI: 10.1145/1669112.1669145.

[186] John Kim, James Balfour, and William Dally. Flattened butterfly topology for on-chip networks. In *Proc. of the 40th International Symposium on Microarchitecture*, pages 172–182, December 2007. DOI: 10.1109/micro.2007.29.

[187] John Kim, William Dally, Steve Scott, and Dennis Abts. Technology-driven, highly-scalable dragonfly topology. In *Proc. of the International Symposium on Computer Architecture*, pages 194–205, June 2008. DOI: 10.1109/isca.2008.19.

[188] Jongman Kim, Chrysostomos Nicopoulos, Dongkook Park, Reetuparna Das, Yuan Xie, N. Vijaykrishnan, Mazin S. Yousif, and Chita R. Das. A novel dimensionally-decomposed router for on-chip communication in 3d architectures. In *International Symposium on Computer Architecture*, pages 138–149, June 2007. DOI: 10.1145/1250662.1250680.

[189] Jongman Kim, Chrysostomos Nicopoulos, Dongkook Park, N. Vijaykrishnan, Mazin S. Yousif, and Chita R. Das. A gracefully degrading and energy-efficient modular router architecture for on-chip networks. In *Proc. of the International Symposium on Computer Architecture*, pages 4–15, June 2006. DOI: 10.1109/isca.2006.6.

[190] Jongman Kim, Dongkook Park, T. Theocharides, N. Vijaykrishnan, and Chita R. Das. A low latency router supporting adaptivity for on-chip interconnects. In *International Conference on Design Automation*, pages 559–564, 2005. DOI: 10.1109/dac.2005.193873.

[191] Joo-Young Kim, Junyoung Park, Seungjin Lee, Minsu Kim, Jinwook Oh, and Hoi-Jun Yoo. A 118.4 GB/s Multi-Casting Network-on-Chip With Hierarchical Star-Ring Combined Topology for Real-Time Object Recognition. *Journal of Solid-State Circuits*, 45(7):1399–1409, 2010. DOI: 10.1109/jssc.2010.2048085.

[192] Ryan Gary Kim, Wonje Choi, Guangshuo Liu, Ehsan Mohandesi, Partha Pratim Pande, Diana Marculescu, and Radu Marculescu. Wireless NoC for VFI-enabled multicore chip design: Performance evaluation and design trade-offs. *IEEE Transactions on Computers*, 65(4):1323–1336, 2016. DOI: 10.1109/tc.2015.2441721.

[193] Michel Kinsy, Myong Hyon Cho, Tina Wen, Edward Suh, Marten van Dijk, and Srinivas Devadas. Application-aware deadlock-free oblivious routing. In *Proc. of the International Symposium on Computer Architecture*, June 2009. DOI: 10.1145/1555754.1555782.

[194] Nevin Kirman, Meyrem Kirman, Rajeev K. Dokania, Jose F. Martinez, Alyssa B. Apsel, Matthew A. Watkins, and David H. Albonesi. Leveraging optical technology in future bus-based chip multiprocessors. In *Proc. of the International Symposium on Microarchitecture*, pages 492–503, December 2006. DOI: 10.1109/micro.2006.28.

[195] Michael Kistler, Michael Perrone, and Fabrizio Petrini. Cell multiprocessor communication network: Built for speed. *IEEE Micro*, 26(3):10–23, May 2006. DOI: 10.1109/mm.2006.49.

[196] Michihiro Koibuchi, Hiroki Matsutani, Hideharu Amano, and Timothy Mark Pinkston. A lightweight fault-tolerant mechanism for network-on-chip. In *Proc. of the Second ACM/IEEE International Symposium on Networks-on-Chip*, NOCS '08, pages 13–22, Washington, DC, USA, 2008. IEEE Computer Society. DOI: 10.1109/nocs.2008.4492721.

[197] Pranay Koka, Michael O. McCracken, Herb Schwetman, Chia-Hsin Chen, Xuezhe Zheng, Ron Ho, Kannan Raj, and Ashok V. Krishnamoorthy. A micro-architectural analysis of switched photonic multi-chip interconnects. In *International Symposium on Computer Architecture*, 2012. DOI: 10.1109/isca.2012.6237014.

[198] Pranay Koka, Michael O. McCracken, Herb Schwetman, Xuezhe Zheng, Ron Ho, and Ashok V. Krishnamoorthy. Silicon-photonic network architectures for scalable, power-efficient multi-chip systems. In *International Symposium on Computer Architecture*, 2010. DOI: 10.1145/1815961.1815977.

[199] Poonacha Kongetira, Kathirgamar Aingaran, and Kunle Olukotun. Niagara: A 32-way multithreaded SPARC processor. *IEEE Micro*, 25(2):21–29, 2005. DOI: 10.1109/mm.2005.35.

[200] Rajesh Kota. HORUS: Large scale SMP using AMD Opteron™. http://www.hypertransport.org/docs/tech/horus_external_white_paper_final.pdf. DOI: 10.1109/MM.2005.28.

[201] Yana Krasteva, Francisco Criado, Eduardo de la Torre, and Teresa Riesgo. A fast emulation-based NoC prototyping framework. In *RECONFIG*, 2008. DOI: 10.1109/reconfig.2008.74.

[202] Yana E. Krasteva, Francisco Criado, Eduardo de la Torre, and Teresa Riesgo. A fast emulation-based NoC prototyping framework. In *Proc. of the 2008 International Conference on Reconfigurable Computing and FPGAs*, RECONFIG '08, pages 211–216, Washington, DC, USA, 2008. IEEE Computer Society. DOI: 10.1109/reconfig.2008.74.

[203] Tushar Krishna. garnet2.0. http://synergy.ece.gatech.edu/tools/garnet/.

[204] Tushar Krishna, Chia-Hsin Owen Chen, Woo Cheol Kwon, and Li-Shiuan Peh. Breaking the on-chip latency barrier using SMART. In *Proc. of the International Symposium on High Performance Computer Architecture*, 2013. DOI: 10.1109/hpca.2013.6522334.

[205] Tushar Krishna, Amit Kumar, Patrick Chiang, Mattan Erez, and Li-Shiuan Peh. NoC with near-ideal express virtual channels using global-line communication. In *Proc. of Hot Interconnects*, pages 11–20, August 2008. DOI: 10.1109/hoti.2008.22.

[206] Tushar Krishna, Li-Shiuan Peh, Bradford M. Beckmann, and Steven K. Reinhardt. Towards the ideal on-chip fabric for 1-to-many and many-to-1 communication. In *Proc. of the International Symposium on Microarchitecture*, December 2011. DOI: 10.1145/2155620.2155630.

[207] John Kubiatowicz and Anant Agarwal. The anatomy of a message in the Alewife multiprocessor. In *Proc. of the International Conference on Supercomputing*, pages 195–206, July 1993. DOI: 10.1145/2591635.2667168.

[208] Amit Kumar, Partha Kundu, Arvind Singh, Li-Shiuan Peh, and Niraj K. Jha. A 4.6Tbits/s 3.6GHz single-cycle NoC router with a novel switch allocator in 65nm CMOS. In *Proc. of the International Conference on Computer Design*, pages 63–70, October 2007. DOI: 10.1109/iccd.2007.4601881.

[209] Amit Kumar, Li-Shiuan Peh, and Niraj K Jha. Token flow control. In *Proc. of the 41st International Symposium on Microarchitecture*, pages 342–353, Lake Como, Italy, November 2008. DOI: 10.1109/micro.2008.4771803.

[210] Amit Kumar, Li-Shiuan Peh, Partha Kundu, and Niraj K. Jha. Express virtual channels: Toward the ideal interconnection fabric. In *Proc. of 34th Annual International Symposium on Computer Architecture*, pages 150–161, San Diego, CA, June 2007. DOI: 10.1145/1250662.1250681.

[211] Shashi Kumar, Axel Jantsch, Juha-Pekka Soininen, M. Forsell, Mikael Millberg, Johnny Öberg, Kari Tiensyrjä, and Ahmed Hemani. A network on chip architecture and design methodology. In *Proc. of the IEEE Computer Society Annual Symposium on VLSI*, pages 105–112, April 2002. DOI: 10.1109/isvlsi.2002.1016885.

[212] G. Kurian, J. Miller, J. Psota, J. Michel, L. Kimerling, and A. Agarwal. ATAC: A 1000-core cache-coherent processor with on-chip optical network. In *International Conference on Parallel Architectures and Compiler Techniques*, 2010. DOI: 10.1145/1854273.1854332.

[213] Hyoukjun Kwon and Tushar Krishna. OpenSMART: Single-cycle multi-hop noc generator in bsv and chisel. In *IEEE International Symposium on Performance Analysis of Systems and Software, ISPASS*, 2017.

[214] Ying-Cherng Lan, Shih-Hsin Lo, , Yueh-Chi Lin, Yu-Hen Hu, and Sao-Jie Chen. BiNoC: A bidirectional NoC architecture with dynamic self-reconfigurable channel. In *Proc. of the International Symposium on Networks-on-Chip*, 2009. DOI: 10.1109/nocs.2009.5071476.

[215] James Laudon and Daniel Lenoski. The SGI Origin: a ccNUMA highly scalable server. In *Proc. of the 24th Annual International Symposium on Computer Architecture*, pages 241–251, May 1997. DOI: 10.1145/264107.264206.

[216] Doowon Lee, Ritesh Parikh, and Valeria Bertacco. Brisk and limited-impact NoC routing reconfiguration. In *Proc. of the Conference on Design, Automation & Test in Europe*, DATE '14, pages 306:1–306:6, 2014. DOI: 10.7873/date2014.319.

[217] Jae W. Lee, Man Cheuk Ng, and Krste Asanović. Globally synchronized frames for guaranteed quality of service in on-chip networks. In *Proc. of the International Symposium on Computer Architecture*, June 2008. DOI: 10.1109/isca.2008.31.

[218] Kangmin Lee, Se-Joong Lee, Sung-Eun Kim, Hye-Mi Choi, Donghyun Kim, Sunyoung Kim, Min-Wuk Lee, and Hoi-Jun Yoo. A 51mW 1.6GHz on-chip network for low-power heterogeneous SoC platform. In *Proc. of the International Solid-State Circuits Conference*, pages 152–153, February 2004. DOI: 10.1109/isscc.2004.1332639.

[219] Kangmin Lee, Se-Joong Lee, and Hoi-Jun Yoo. Low-power network-on-chip for high-performance SoC design. *IEEE Transactions on VLSI Systems*, 14(2), February 2006. DOI: 10.1109/tvlsi.2005.863753.

[220] Michael Lee, John Kim, Dennis Abts, Mike Marty, and Jae Lee. Probabilistic distance-based arbitration: Providing equality of service for many-core cmps. In *Proc. of the International Symposium on Microarchitecture*, 2010. DOI: 10.1109/micro.2010.18.

[221] Whay Sing Lee, William J. Dally, Stephen W. Keckler, Nicholas P. Carter, and Andrew Chang. An efficient protected message interface in the MIT M-Machine. *IEEE Computer Special Issue on Design Challenges for High Performance Network Interfaces*, 31(11):69–75, November 1998. DOI: 10.1109/2.730739.

[222] Charles Leiserson. Fat-trees: Universal networks for hardware-efficient supercomputing. *IEEE Transactions on Computers*, 34(10):892–901, October 1985. DOI: 10.1109/tc.1985.6312192.

[223] Feihui Li, Chrysostomos Nicopoulos, Thomas Richardson, Yuan Xie, N. Vijaykrishnan, and Mahmut Kandemir. Design and management of 3D chip multiprocessors using network-in-memory. In *Proc. of the International Symposium on Computer Architecture*, pages 130–141, June 2006. DOI: 10.1109/isca.2006.18.

[224] Sheng Li, Jung Ho Ahn, Richard D. Strong, Jay B. Brockman, Dean M. Tullsen, and Norman P. Jouppi. McPAT: an integrated power, area, and timing modeling framework for multicore and manycore architectures. In *42st Annual IEEE/ACM International Symposium on Microarchitecture (MICRO)*, pages 469–480, 2009. DOI: 10.1145/1669112.1669172.

[225] Zheng Li, Jie Wu, Li Shang, Robert Dick, and Yihe Sun. Latency criticality aware on-chip communication. In *Proc. of the IEEE Conference on Design, Automation, and Test in Europe*, March 2009. DOI: 10.1109/date.2009.5090820.

[226] Zimo Li, Joshua San Miguel, and Natalie Enright Jerger. The runahead network-on-chip. In *International Symposium on High Performance Computer Architecture*, 2016. DOI: 10.1109/hpca.2016.7446076.

[227] Erik Lindholm, John Nickolls, Stuart Oberman, and John Montrym. NVIDIA Tesla: A unified graphics and computing architecture. *IEEE Micro*, 28(2):39–55, March-April 2008. DOI: 10.1109/mm.2008.31.

[228] Jifeng Liu, Lionel C Kimerling, and Jurgen Michel. Monolithic Ge-on-Si lasers for large-scale electronic-photonic integration. *Semiconductor Science and Technology*, 27(9):094006, 2012. DOI: 10.1088/0268-1242/27/9/094006.

[229] Yong Liu, Ping-Hsuan Hsieh, Seongwon Kim, Jae-sun Seo, Robert Montoye, Leland Chang, Jose Tierno, and Daniel Friedman. A 0.1 pJ/b 5-to-10Gb/s charge-recycling stacked low-power I/O for on-chip signaling in 45nm CMOS SOI. In *2013 IEEE International Solid-State Circuits Conference Digest of Technical Papers*, pages 400–401. IEEE, 2013. DOI: 10.1109/isscc.2013.6487787.

[230] Pejman Lotfi-Kamran, Boris Grot, and Babak Falsafi. NOC-Out: Microarchitecting a scale-out processor. In *Proc. of the International Symposium on Microarchitecture*, 2012. DOI: 10.1109/micro.2012.25.

[231] Zhonghai Lu, Ming Lui, and Axel Jantsch. Layered switching for networks on chip. In *Proc. of the Conference on Design Automation*, pages 122–127, San Diego, CA, June 2007. DOI: 10.1109/dac.2007.375137.

[232] Lian-Wee Luo, Noam Ophir, Christine P. Chen, Lucas H. Gabrielli, Carl B. Poitras, Keren Bergmen, and Michal Lipson. WDM-compatible mode-division multiplexing on a silicon chip. *Nature Communications*, 5:3069 EP –, 01 2014. DOI: 10.1038/ncomms4069.

[233] Sheng Ma, Natalie Enright Jerger, and Zhiying Wang. DBAR: an efficient routing algorithm to support multiple concurrent applications in networks-on-chip. In *Proc. of the International Symposium on Computer Architecture*, June 2011. DOI: 10.1145/2000064.2000113.

[234] Sheng Ma, Natalie Enright Jerger, and Zhiying Wang. Supporting efficient collective communication in NoCs. In *Proc. of the International Symposium on High Performance Computer Architecture*, February 2012. DOI: 10.1109/hpca.2012.6168953.

[235] Sheng Ma, Natalie Enright Jerger, and Zhiying Wang. Whole packet forwarding: Efficient design of fully adaptive routing algorithms for networks-on-chip. In *Proc. of the International Symposium on High Performance Computer Architecture*, February 2012. DOI: 10.1109/hpca.2012.6169049.

[236] Sheng Ma, Natalie Enright Jerger, Zhiying Wang, Ming-Che Lai, and Libo Huang. Holistic routing algorithm design to support workload consolidation in NoCs. *IEEE Transactions on Computers*, 63(3), March 2014. DOI: 10.1109/tc.2012.201.

[237] Sheng Ma, Zhiying Wang, Zonglin Liu, and Natalie Enright Jerger. Leaving one slot empty: Flit bubble flow control for torus cache-coherent NoCs. *IEEE Transactions on Computers*, 64:763–777, March 2015. DOI: 10.1109/tc.2013.2295523.

[238] Radu Marculescu, Umit Y. Ogras, Li-Shiuan Peh, Natalie Enright Jerger, and Yatin Hoskote. Outstanding research problems in NoC design: System, microarchitecture, and circuit perspectives. *IEEE Transactions on Computer-Aided Design of Integrated Circuits and Systems*, 28(1):3–21, January 2009. DOI: 10.1109/tcad.2008.2010691.

[239] Theodore Marescaux, Andrei Bartic, Diderick Verkest, D. Verkest, Rudy Lauwereins, Serge Vernalde, and R. Lauwereins. Interconnection networks enable fine-grain dynamic multi-tasking on FPGAs. In *International Conference on Field-Programmable Logic and Applications*, 2002. DOI: 10.1007/3-540-46117-5_82.

[240] Michael R. Marty and Mark D. Hill. Coherence ordering for ring-based chip multiprocessors. In *Proc. of the 39th International Symposium on Microarchitecture*, pages 309–320, December 2006. DOI: 10.1109/micro.2006.14.

[241] Michael R. Marty and Mark D. Hill. Virtual hierarchies to support server consolidation. In *Proc. of the International Symposium on Computer Architecture*, pages 46–56, June 2007. DOI: 10.1145/1250662.1250670.

[242] Hiroki Matsutani, Michihiro Koibuchi, Hideharu Amano, and Tsutomu Yoshinaga. Prediction router: Yet another low latency on-chip router architecture. In *Proc. of the International Symposium on High Performance Computer Architecture*, pages 367–378, February 2009. DOI: 10.1109/hpca.2009.4798274.

[243] George Michelogiannakis, James Balfour, and William J. Dally. Elastic-buffer flow control for on-chip networks. In *Proc. of the International Symposium on High Performance Computer Architecture*, pages 151–162, February 2009. DOI: 10.1109/hpca.2009.4798250.

[244] George Michelogiannakis, Nan Jiang, Daniel Becker, and William J. Dally. Packet chaining: Efficient single-cycle allocation for on-chip networks. In *Proc. of the International Symposium on Microarchitecture*, 2011. DOI: 10.1145/2155620.2155631.

[245] ST Microelectronics. http://www.st.com.

[246] ST Microelectronics. STBus interconnect. http://www.st.com/stonline/products/technologies/soc/stbus.htm.

[247] Mikael Millberg, Erland Nilsson, Rikard Thid, and Axel Jantsch. Guaranteed bandwidth using looped containers in temporally disjoint networks within the Nostrum network-on-chip. In *Proc. of the conference on Design, Automation and Testing in Europe (DATE)*, pages 890–895, 2004. DOI: 10.1109/date.2004.1269001.

[248] D.A.B. Miller. Rationale and challenges for optical interconnects to electronic chips. *Proc. of the IEEE*, 88(6):728–749, June 2000. DOI: 10.1109/5.867687.

[249] David Miller. Attojoule optoelectronics for low-energy information processing and communications: a tutorial review. https://arxiv.org/abs/1609.05510v2. DOI: 10.1109/jlt.2017.2647779.

[250] Asit K Mishra, Reetuparna Das, Soumya Eachempati, Ravi Iyer, Narayanan Vijaykrishnan, and Chita R Das. A case for dynamic frequency tuning in on-chip networks. In *2009 42nd Annual IEEE/ACM International Symposium on Microarchitecture (MICRO)*, pages 292–303. IEEE, 2009. DOI: 10.1145/1669112.1669151.

[251] MIT-DSENT. MIT-DSENT. https://sites.google.com/site/mitdsent/.

[252] Thomas Moscibroda and Onur Mutlu. A case for bufferless routing in on-chip networks. In *Proc. of the 36th International Symposium on Computer Architecture*, June 2009. DOI: 10.1145/1555754.1555781.

[253] Thomas Moscibroda and Onur Mutlu. A case for bufferless routing in on-chip networks. In *36th International Symposium on Computer Architecture (ISCA)*, pages 196–207, Austin, TX, June 20–24, 2009. DOI: 10.1145/1555754.1555781.

[254] Shubhendu S. Mukherjee, Peter Bannon, Steven Lang, Aaron Spink, and David Webb. The Alpha 21364 network architecture. *IEEE Micro*, 22(1):26–35, 2002. DOI: 10.1109/40.988687.

[255] Robert Mullins, Andrew West, and Simon Moore. Low-latency virtual-channel routers for on-chip networks. In *Proc. of the International Symposium on Computer Architecture*, pages 188–197, June 2004. DOI: 10.1109/isca.2004.1310774.

[256] S. Murali and Giovanni De Micheli. SUNMAP: A tool for automatic topology selection and generation for NoCs. In *Proc. of the Design Automation Conference*, pages 914–919, June 2004. DOI: 10.1145/996566.996809.

[257] Srinivasan Murali and Giovanni De Micheli. Bandwidth-constrained mapping of cores onto NoC architectures. In *Proc. of the Conference for Design, Automation and Test in Europe*, pages 896–901, February 2004. DOI: 10.1109/date.2004.1269002.

[258] Srinivasan Murali, Paolo Meloni, Federico Angiolini, David Atienza, Salvatore Carta, Luca Benini, Giovanni De Micheli, and Luigi Raffo. Designing application-specific networks on chips with floorplan information. In *International Conference on Computer-Aided Design*, pages 355–362, November 2006. DOI: 10.1109/iccad.2006.320058.

[259] Ted Nesson and S. Lennart Johnsson. ROMM routing on mesh and torus networks. In *Proc. of the Symposium on Parallel Algorithms and Architectures*, pages 275–287, 1995. DOI: 10.1145/215399.215455.

[260] Chrysostomos A. Nicopoulos, Dongkook Park, Jongman Kim, N. Vijaykrishnan, Mazin S. Yousif, and Chita R. Das. ViChaR: A dynamic virtual channel regulator for network on-chip routers. In *International Symposium on Microarchitecture*, pages 333–344, December 2006. DOI: 10.1109/micro.2006.50.

[261] Rishiyur Nikhil. Bluespec system verilog: efficient, correct rtl from high level specifications. In *MEMOCODE*, pages 69–70. IEEE, 2004. DOI: 10.1109/memcod.2004.1459818.

[262] Erland Nilsson, Mikael Millberg, Johnny Oberg, and Axel Jantsch. Load distribution with proximity congestion awareness in a network on chip. In *Proc. of the Conference on Design, Automation and Test in Europe*, pages 1126–1127, March 2003. DOI: 10.1109/date.2003.1253765.

[263] Christopher Nitta, Kevin Macdonald, Matthew Farrens, and Venkatesh Akella. Inferring packet dependencies to improve trace based simulation of on-chip networks. In *International Symposium on Networks on Chip*, 2011. DOI: 10.1145/1999946.1999971.

[264] CMU-SAFARI NOCulator. Noculator. https://github.com/CMU-SAFARI/NOCulator.

[265] Peter R. Nuth and William J. Dally. The J-Machine network. In *Proc. of the International Conference on Computer Design*, pages 420–423, October 1992. DOI: 10.1109/iccd.1992.276305.

[266] Umit Y Ogras, Paul Bogdan, and Radu Marculescu. An analytical approach for network-on-chip performance analysis. *IEEE Transactions on Computer-Aided Design of Integrated Circuits and Systems*, 29(12):2001–2013, 2010. DOI: 10.1109/tcad.2010.2061613.

[267] Umit Y. Ogras, Jingcao Hu, and Radu Marculescu. Key research problems in NoC design: A holistic perspective. In *Proc. of the International Conference on Hardware-Software Codesign Systems and Synthesis*, pages 69–74, September 2005. DOI: 10.1145/1084834.1084856.

[268] Umit Y. Ogras and Radu Marculescu. "It's a small world after all": NoC performance optimization via long-range link insertion. *IEEE Transactions on Very Large Scale Integration (VLSI) Systems - Special Section Hardware/Software Codesign System Synthesis*, 14(7):693–706, July 2006. DOI: 10.1109/tvlsi.2006.878263.

[269] Umit Y Ogras, Radu Marculescu, Diana Marculescu, and Eun Gu Jung. Design and management of voltage-frequency island partitioned networks-on-chip. *IEEE Transactions on Very Large Scale Integration (VLSI) Systems*, 17(3):330–341, 2009. DOI: 10.1109/tvlsi.2008.2011229.

[270] Jungju Oh, Milos Prvulovic, and Alenka Zajic. TLSync: Support for multiple fast barriers using on-chip transmission lines. In *Proc. of the 38th Annual International Symposium on Computer Architecture*, ISCA '11, pages 105–116, 2011. DOI: 10.1145/2000064.2000078.

[271] Kunle Olukotun, Basem A. Nayfeh, Lance Hammond, Ken Wilson, and Kun-Yung Chang. The case for a single-chip multiprocessor. In *Proc. of the International Symposium on Architectural Support for Parallel Languages and Operating Systems*, pages 2–11, October 1996. DOI: 10.1145/237090.237140.

[272] Network on Chip (NoC) Blog. Network-on-chip (noc) blog. https://networkonchi p.wordpress.com/2011/02/22/simulators/.

[273] Marta Ortin, Dario Suarez, Maria Villarroya, Cruz Izu, and Victor Vinals. Dynamic construction of circuits for reactive traffic in homogeneous CMPs. In *Design, Automation and Test in Europe Conference and Exhibition (DATE), 2014*, pages 1–4. IEEE, 2014. DOI: 10.7873/date.2014.254.

[274] Ralph H. Otten and Robert K. Brayton. Planning for performance. In *Proc. of the conference on Design Automation*, pages 122–127, June 1998. DOI: 10.1109/dac.1998.724452.

[275] Jin Ouyang and Yuan Xie. LOFT: A high performance network-on-chip providing quality of service support. In *Proc. of the International Symposium on Microarchitecture*, 2010. DOI: 10.1109/micro.2010.21.

[276] John D. Owens, William J. Dally, Ron Ho, D. N. Jayasimha, Stephen W. Keckler, and Li-Shiuan Peh. Research challanges for on-chip interconnection networks. *IEEE Micro, Special Issue on On-Chip Interconnects for Multicores*, 27(5):96–108, September/October 2007. DOI: 10.1109/mm.2007.4378787.

[277] Hewlett Packard. Supersim. https://github.com/hewlettpackard/supersim.

[278] Maurizio Palesi, Rickard Holsmark, Shashi Kumar, and Vincenzo Catania. A methodology for design of application specific deadlock-free routing algorithms for NoC systems. In *Proc. of the International Conference on Hardware-Software Codesign Systems and Synthesis*, pages 142–147, October 2006. DOI: 10.1145/1176254.1176289.

[279] Yan Pan, John Kim, and Gokhan Memik. FlexiShare: Energy-efficient nanophotonic crossbar architecture through channel sharing. In *International Symposium on High Performance Computer Architecture*, 2010. DOI: 10.1109/HPCA.2010.5416626.

[280] Yan Pan, Prabhat Kumar, John Kim, Gokhan Memik, Yu Zhang, and Alok Choudhary. Firefly: Illuminating future network-on-chip with nanophotonics. In *Proc. of the International Symposium on Computer Architecture*, June 2009. DOI: 10.1145/1555754.1555808.

[281] Michael K. Papamichael and James C. Hoe. Connect: re-examining conventional wisdom for designing NoCs in the context of FPGAs. In *Proc. of the ACM/SIGDA international symposium on Field Programmable Gate Arrays*, pages 37–46. ACM, 2012. DOI: 10.1145/2145694.2145703.

[282] Ritesh Parikh and Valeria Bertacco. uDIREC: Unified diagnosis and reconfiguration for frugal bypass of NoC faults. In *International Symposium on Microarchitecture*, 2013. DOI: 10.1145/2540708.2540722.

[283] Ritesh Parikh, Reetuparna Das, and Valeria Bertacco. Power-aware NoCs through routing and topology reconfiguration. In *2014 51st ACM/EDAC/IEEE Design Automation Conference (DAC)*, pages 1–6. IEEE, 2014. DOI: 10.1109/dac.2014.6881489.

[284] Ritesh Parikh, Rawan Abdel Khalek, and Valeria Bertacco. Formally enhanced runtime verification to ensure NoC functional correctness. In *International Symposium on Microarchitecture*, 2011. DOI: 10.1145/2155620.2155668.

[285] Dongkook Park, Reetuparna Das, Chrysostomos Nicopoulos, Jongman Kim, N. Vijaykrishnan, Ravishankar Iyer, and Chita R. Das. Design of a dynamic priority-based fast path architecture for on-chip interconnects. In *Proc. of the 15th IEEE Symposium on High-Performance Interconnects*, pages 15–20, August 2007. DOI: 10.1109/hoti.2007.1.

[286] Dongkook Park, Soumya Eachempati, Reetuparna Das, Asit K. Mishra, Yuan Xie, N. Vijaykrishnan, and Chita R. Das. MIRA: A multi-layered on-chip interconnect router architecture. In *Proc. of the International Symposium on Computer Architecture*, pages 251–261, June 2008. DOI: 10.1109/isca.2008.13.

[287] S. Park, T. Krishna, C.-H. O. Chen, B. Daya, A. P. Chandrakasan, and L.-S. Peh. Approaching the Theoretical Limits of a Mesh NoC with a 16-Node Chip Prototype in 45nm SOI. In *Proc. of the ACM/EDAC/IEEE Design Automation Conference (DAC)*, pages 398–405, 2012. DOI: 10.1145/2228360.2228431.

[288] Sunghyun Park, Masood Qazi, Li-Shiuan Peh, and Anantha P. Chandrakasan. 40.4fJ/bit/mm low-swing on-chip signaling with self-resetting logic repeaters embedded within a mesh NoC in 45nm SOI CMOS. In *Design, Automation and Test in Europe (DATE)*, pages 1637–1642, 2013. DOI: 10.7873/date.2013.332.

[289] Sudeep Pasricha and Nikil Dutt. *On-Chip Communication Architectures: System on Chip Interconnect*. Morgan Kaufmann, 2008.

[290] Li-Shiuan Peh and William J. Dally. Flit-reservation flow control. In *Proc. of the 6th International Symposium on High Performance Computer Architecture*, pages 73–84, February 2000. DOI: 10.1109/hpca.2000.824340.

[291] Li-Shiuan Peh and William J. Dally. A delay model and speculative architecture for pipelined routers. In *Proc. of the International Symposium on High Performance Computer Architecture*, pages 255–266, January 2001. DOI: 10.1109/hpca.2001.903268.

[292] Li-Shiuan Peh and William J. Dally. A delay model for router microarchitectures. *IEEE Micro*, 21(1):26–34, January 2001. DOI: 10.1109/40.903059.

[293] D. Pham, T. Aipperspach, D. Boerstler, M. Bolliger, R. Chaudhry, D. Cox, P. Harvey, H. P. Hofstee, C. Johns, J. Kahle, A. Kameyama, J. Keaty, Y. Masubuchi, M. Pham, J. Pille,

S. Posluszny, M. Riley, D. Stasiak, M. Suzuoki, O. Takahashi, J. Warnock, S. Weitzel, D. Wendel, and K. Yazawa. Overview of the architecture, circuit design, and physical implementation of a first-generation cell processor. *IEEE Journal of Solid-State Circuits*, 41(1):179–196, 2006. DOI: 10.1109/jssc.2005.859896.

[294] Timothy Mark Pinkston and José Duato. Appendix F: Interconnection networks. In John L. Hennessy and David A. Patterson, Eds., *Computer Architecture: A Quantitative Approach*, pages 1–114. Elsevier Publishers, 5th edition, September 2011.

[295] Alessandro Pinto, Luca P. Carloni, and Alberto L. Sangiovanni-Vincentelli. Efficient synthesis of networks on chip. In *Proc. of the International Conference on Computer Design*, pages 146–150, October 2003. DOI: 10.1109/iccd.2003.1240887.

[296] Andreas Prodromou, Andreas Panteli, Chrysostomos Nicopoulos, and Yiannakis Sazeides. Nocalert: An on-line and real-time fault detection mechanism for network-on-chip architectures. In *International Symposium on Microarchitecture*, 2012. DOI: 10.1109/micro.2012.15.

[297] Open Core Protocol. http://www.ocpip.org/home/.

[298] V. Puente, C. Izu, R. Beivide, J. A. Gregorio, F. Vallejo, and J. M. Prellezo. The adaptive bubble router. *Journal of Parallel and Distributed Computing*, 64(9):1180–1208, 2001. DOI: 10.1006/jpdc.2001.1746.

[299] Antonio Pullini, Federico Angiolini, Davide Bertozzi, and Luca Benini. Fault tolerance overhead in network-on-chip flow control schemes. In *Proc. of the Symposium on Integrated and Circuits System Design*, pages 224–229, Sept 2005. DOI: 10.1109/sbcci.2005.4286861.

[300] Martin Radetzki, Chaochao Feng, Xueqian Zhao, and Axel Jantsch. Methods for fault tolerance in networks-on-chip. *ACM Computing Surveys*, 46(1):8:1–8:38, July 2013. DOI: 10.1145/2522968.2522976.

[301] Mukund Ramakrishna, Paul Gratz, and Alexander Sprintson. GCA: Global congestion awareness for load balance in networks-on-chip. In *Proc. of the International Symposium on Networks-on-Chip*, April 2013. DOI: 10.1109/nocs.2013.6558405.

[302] Aniruddh Ramrakhyani and Tushar Krishna. Static bubble: A framework for deadlock-free irregular on-chip topologies. In *Proc. of the International Symposium on High Performance Computer Architecture*, 2017.

[303] José Renau, B. Fraguela, James Tuck, Wei Liu, Milos Prvulovic, Luis Ceze, Karin Strauss, Smruti Sarangi, Paul Sack, and Pablo Montesinos. SESC simulator. http://sesc.sourceforge.net.

[304] Samuel Rodrigo, Jose Flich, José Duato, and Mark Hummel. Efficient unicast and multicast support for CMPs. In *Proc. of the International Symposium on Microarchitecture*, pages 364–375, November 2008. DOI: 10.1109/micro.2008.4771805.

[305] Manuel Saldana, Lesley Shannon, and Paul Chow. The routability of multiprocessor network topologies in FPGAs. In *International Workshop on System-level Interconnect Prediction*, 2006. DOI: 10.1145/1117278.1117290.

[306] Ahmad Samih, Ren Wang, Anil Krishna, Christian Maciocco, Charlie Tai, and Yan Solihin. Energy-efficient interconnect via router parking. In *High Performance Computer Architecture (HPCA2013), 2013 IEEE 19th International Symposium on*, pages 508–519. IEEE, 2013. DOI: 10.1109/hpca.2013.6522345.

[307] Daniel Sanchez and Christos Kozyrakis. ZSim: fast and accurate microarchitectural simulation of thousand-core systems. In *The 40th Annual International Symposium on Computer Architecture, ISCA'13*, pages 475–486, Tel-Aviv, Israel, June 23–27, 2013. DOI: 10.1145/2508148.2485963.

[308] Karthikeyan Sankaralingam, Ramadass Nagarajan, Haiming Liu, Jaehyuk Huh, Changkyu Kim, Doug Burger, Stephen W. Keckler, and Charles R. Moore. Exploiting ILP, TLP, and DLP using polymorphism in the TRIPS architecture. In *Proc. of the 30th Annual International Symposium on Computer Architecture*, pages 422–433, June 2003. DOI: 10.1109/isca.2003.1207019.

[309] Graham Schelle and Dirk Grunwald. On-chip interconnect exploration for multicore processors utilizing FPGAs. In *Proceedings of the 2nd Workshop on Architecture Research using FPGA Platforms*, February 2006.

[310] M. D. Schroeder, A. D. Birrell, M. Burrows, H. Murray, R. M. Needham, T. L. Rodeheffer, E. H. Satterthwaite, and C. P. Thacker. Autonet: A high-speed, self-configuring local area network using point-to-point links. *IEEE Journal on Selected Areas in Communications*, 9(8):1318–1335, September 2006. DOI: 10.1109/49.105178.

[311] Steve Scott, Dennis Abts, John Kim, and William J. Dally. The BlackWidow high-radix Clos network. In *Proc. of the International Symposium on Computer Architecture*, pages 16–27, June 2006. DOI: 10.1109/isca.2006.40.

[312] Larry Seiler, Doug Carmean, Eric Sprangle, Tom Forsyth, Michael Abrash, Pradeep Dubey, Stephen Junkins, Adam Lake, Jeremy Sugerman, Robert Cavin, Roger Espasa, Ed Grochowski, Toni Juan, and Pat Hanrahan. Larrabee: A many-core x86 architecture for visual computing. *ACM Transactions on Graphics*, 27, August 2008. DOI: 10.1145/1360612.1360617.

[313] Daeho Seo, Akif Ali, Won-Taek Lim, Nauman Rafique, and Mithuna Thottenhodi. Near-optimal worst-case throughput routing for two-dimensional mesh networks. In *Proc. of the 32nd Annual International Symposium on Computer Architecture*, pages 432–443, June 2005. DOI: 10.1109/isca.2005.37.

[314] Jae-sun Seo, Ron Ho, Jon K. Lexau, Michael Dayringer, Dennis Sylvester, and David Blaauw. High-bandwidth and low-energy on-chip signaling with adaptive pre-emphasis in 90nm CMOS. In *IEEE International Solid-State Circuits Conference, ISSCC 2010, Digest of Technical Papers*, pages 182–183, San Francisco, CA, February 7–11, 2010. DOI: 10.1109/isscc.2010.5433993.

[315] Korey Sewell, Ronald G. Dreslinski, Thomas Manville, Sudhir Satpathy, Nathaniel Ross Pinckney, Geoffrey Blake, Michael Cieslak, Reetuparna Das, Thomas F. Wenisch, Dennis Sylvester, David Blaauw, and Trevor N. Mudge. Swizzle-Switch Networks for Many-Core Systems. *IEEE Journal on Emergerging and Selected Topics in Circuits And Systems*, 2(2):278–294, 2012. DOI: 10.1109/jetcas.2012.2193936.

[316] Assaf Shacham, Keren Bergman, and Luca P. Carloni. On the design of a photonic network on chip. In *Proc. of International Symposium on Networks-on-Chip*, pages 53–64, May 2007. DOI: 10.1109/nocs.2007.35.

[317] Akbar Sharifi, Emre Kultursay, Mahmut Kandemir, and Chita R. Das. Addressing end-to-end memory access latency in NoC-based multicores. In *Proc. of the International Symposium on Microarchitecture*, 2012. DOI: 10.1109/micro.2012.35.

[318] Keun Sup Shim, Myong Hyon Cho, Michel Kinsy, Tina Wen, Mieszko Lis, G Edward Suh, and Srinivas Devadas. Static virtual channel allocation in oblivious routing. In *Proc. of the 2009 3rd ACM/IEEE International Symposium on Networks-on-Chip*, pages 38–43. IEEE Computer Society, 2009. DOI: 10.1109/nocs.2009.5071443.

[319] Arjun Singh, William J. Dally, Amit K. Gupta, and Brian Towles. GOAL: A load-balanced adaptive routing algorithm for torus networks. In *Proc. of the International Symposium on Computer Architecture*, pages 194–205, June 2003. DOI: 10.1109/isca.2003.1207000.

[320] Arjun Singh, William J. Dally, Brian Towles, and Amit K. Gupta. Locality-preserving randomized oblivious routing on torus networks. In *SPAA*, pages 9–13, 2002. DOI: 10.1145/564870.564873.

[321] Avinash Sodani, Roger Gramunt, Jesus Corbal, Ho-Seop Kim, Krishna Vinod, Sundaram Chinthamani, Steven Hutsell, Rajat Agarwal, and Yen-Chen Liu. Knights Landing: Second-generation Intel Xeon Phi product. *IEEE Micro*, 36(2):34–46, 2016. DOI: 10.1109/hotchips.2015.7477467.

[322] Yong Ho Song and Timothy Mark Pinkston. A progressive approach to handling message-dependent deadlocks in parallel computer systems. *IEEE Transactions on Parallel and Distributed Systems*, 14(3):259–275, March 2003. DOI: 10.1109/tpds.2003.1189584.

[323] Sonics. SonicsGN. `http://sonicsinc.com/products/on-chip-networks/sonicsgn/`.

[324] Sonics Inc. `http://www.sonicsinc.com/home/htm`.

[325] Daniel J. Sorin, Mark D. Hill, and David A. Wood. *A Primer on Memory Consistency and Cache Coherence*. Morgan Claypool, 2011. DOI: 10.2200/s00346ed1v01y201104cac016.

[326] Krishnan Srinivasan and Karam S. Chatha. A low complexity heuristic for design of custom network-on-chip architectures. In *Proc. of the Conference on Design, Automation and Test in Europe*, pages 130–135, 2006. DOI: 10.1109/date.2006.244034.

[327] S. Stergiou, E Angiolini, D. Bertozzi, S. Carta, L. Raffo, and G. De Micheli. xpipesLite: A synthesis-oriented design flow for networks on chip. In *Proc. of the Conference on Design, Automation and Test Europe*, pages 1188–1193, 2005. DOI: 10.1109/date.2005.1.

[328] C. Sun, C.-H. O. Chen, G. Kurian, L. Wei, J. Miller, A. Agarwal, L.-S. Peh, and V. Stojanovic. DSENT - A Tool Connecting Emerging Photonics with Electronics for Opto-Electronic Networks-on-Chip Modeling. In *Proc. of International Symposium on Networks-on-Chip*, pages 201–210, 2012. DOI: 10.1109/nocs.2012.31.

[329] C. Sun, Mark T. Wade, Yunsup Lee, Jason S. Orcutt, Luca Alloatti, Michael S. Georgas, Andrew S. Waterman, Jeffrey M. Shainline, Rimas R. Avizienis, Sen Lin, Benjamin R. Moss, Rajesh Kumar, Fabio Pavanello, Amir H. Atabaki, Henry M. Cook, Albert J. Ou, Jonathan C. Leu, Yu-Hsin Chen, Krste Asanovic, Rajeev J. Ram, Milos A. Popovic, and Vladimir M. Stojanovic. Single-chip microprocessor that communicates directly using light. *Nature*, 528(7583):534–538, 2015. DOI: 10.1038/nature16454.

[330] Chen Sun, Mark Wade, Michael Georgas, Sen Lin, Luca Alloatti, Benjamin Moss, Rajesh Kumar, Amir H Atabaki, Fabio Pavanello, Jeffrey M Shainline, Jason S. Orcutt, Rajeev J. Ram, Milos Popovic, and Vladimir Stojanovic. A 45 nm CMOS-SOI monolithic photonics platform with bit-statistics-based resonant microring thermal tuning. *IEEE Journal of Solid-State Circuits*, 51(4):893–907, 2016. DOI: 10.1109/jssc.2016.2519390.

[331] Steven Swanson, Ken Michelson, Andrew Schwerin, and Mark Oskin. Wavescalar. In *Proc. of the 36th International Symposium on Microarchitecture*, pages 291–302, 2003. DOI: 10.1109/micro.2003.1253203.

[332] Yasuhiro Take, Hiroki Matsutani, Daisuke Sasaki, Michihiro Koibuchi, Tadahiro Kuroda, and Hideharu Amano. 3D NoC with inductive-coupling links for building-block SiPs. *IEEE Transactions on Computers*, 63(3):748–763, 2014. DOI: 10.1109/tc.2012.249.

[333] Y. Tamir and H. C. Chi. Symmetric crossbar arbiters for VLSI communication switches. *IEEE Transactions Parallel and Distributed Systems*, 4(1):13–27, 1993. DOI: 10.1109/71.205650.

[334] Yuval Tamir and Gregory L. Frazier. Dynamically-allocated multi-queue buffers for VLSI communication switches. *IEEE Transactions on Computers*, 41(6):725–737, June 1992. DOI: 10.1109/12.144624.

[335] Michael Bedford Taylor, Walter Lee, Saman Amarasinghe, and Anant Agarwal. Scalar operand networks: On-chip interconnect for ILP in partitioned architectures. In *Proc. of the International Symposium on High Performance Computer Architecture*, pages 341–353, February 2003. DOI: 10.1109/hpca.2003.1183551.

[336] J. Tendler, J. Dodson, J.S. Fields, H. Le, and B. Sinharoy. Power4 system microarchitecture. *IBM Journal of Research and Development*, 46(1):5–26, 2002. DOI: 10.1147/rd.461.0005.

[337] Kevin Tien, Noah Sturcken, Naigang Wang, Jae-woong Nah, Bing Dang, Eugene J. O'Sullivan, Paul S. Andry, Michele Petracca, Luca P. Carloni, William J. Gallagher, and Kenneth L. Shepard. An 82%-efficient multiphase voltage-regulator 3D interposer with on-chip magnetic inductors. In *VLSIC*, page 192. IEEE, 2015.

[338] Brian Towles, J. P. Grossman, Brian Greskamp, and David E. Shaw. Unifying on-chip and inter-node switching within the Anton 2 network. In *Proceeding of the 41st Annual International Symposium on Computer Architecuture*, ISCA '14, pages 1–12, Piscataway, NJ, USA, 2014. IEEE Press. DOI: 10.1109/isca.2014.6853238.

[339] Anh Thien Tran, Dean Nguyen Truong, and Bevan M. Baas. A reconfigurable source-synchronous on-chip network for GALS many-core platforms. *IEEE Transactions on CAD of Integrated Circuits and Systems*, 29(6):897–910, 2010. DOI: 10.1109/tcad.2010.2048594.

[340] Marc Tremblay and Shailender Chaudhry. A third-generation 65nm 16-core 32-thread plus 32-scout-thread CMT SPARC processor. In *Proc. of the International Solid-State Circuits Conference*, 2008. DOI: 10.1109/isscc.2008.4523067.

[341] Sebastian Turullols and Ram Sivaramakrishnan. Sparc t5: 16-core cmt processor with glueless 1-hop scaling to 8-sockets. In *Hot Chips 24 Symposium (HCS), 2012 IEEE*, pages 1–37. IEEE, 2012. DOI: 10.1109/hotchips.2012.7476504.

[342] L. G. Valiant and G. J. Brebner. Universal schemes for parallel communication. In *Proc. of the 13th Annual ACM Symposium on Theory of Computing*, pages 263–277, 1981. DOI: 10.1145/800076.802479.

[343] J.W. van den Brand, C. Ciordas, K. Goossens, and T. Basten. Congestion-controlled best-effort communication for networks-on-chip. In *Proc. of the Conference on Design, Automation and Test in Europe*, pages 948–953, April 2007. DOI: 10.1109/date.2007.364415.

[344] S. Vangal, J. Howard, G. Ruhl, S. Dighe, H. Wilson, J. Tschanz, D. Finan, P. Iyer, A. Singh, T. Jacob, S. Jain, S. Venkataraman, Y. Hoskote, and N. Borkar. An 80-tile 1.28 TFLOPS network-on-chip in 65nm CMOS. In *Proc. of the IEEE International Solid-State Circuits Conference (ISSCC)*, pages 98–99, February 2007. DOI: 10.1109/isscc.2007.373606.

[345] Dana Vantrease, Robert Schreiber, Matteo Monchiero, Moray McLaren, Norman P. Jouppi, Marco Fiorentino, Al Davis, Nathan L. Binkert, Raymond G. Beausoleil, and Jung Ho Ahn. Corona: System implications of emerging nanophotonic technology. In *International Symposium on Computer Architecture*, pages 153–164, June 2008. DOI: 10.1109/isca.2008.35.

[346] Anja von Beuningen and Ulf Schlichtmann. PLATON: A Force-Directed Placement Algorithm for 3D Optical Networks-on-Chip. In *International Symposium on Physical Design*. ACM, 2016. DOI: 10.1145/2872334.2872356.

[347] S. Wamakulasuriya and T.M. Pinkston. Characterization of deadlocks in k-ary n-cube networks. *IEEE Transactions on Parallel and Distributed Systems*, 10(9):904–921, 1999. DOI: 10.1109/71.798315.

[348] S. Wamakulasuriya and T.M. Pinkston. A formal model of message blocking and deadlock resolution in interconnection networks. *IEEE Transactions on Parallel and Distributed Systems*, 11(3):212–229, 2000. DOI: 10.1109/71.841739.

[349] Danyao Wang, Natalie Enright Jerger, and J. Gregory Steffan. DART: A programmable architecture for NoC simulation on FPGAs. In *International Network on Chip Symposium (NOCS)*, pages 145–152, 2011. DOI: 10.1145/1999946.1999970.

[350] Danyao Wang, Charles Lo, Jasmina Vasiljevic, Natalie Enright Jerger, and J. Gregory Steffan. DART: A programmable architecture for NoC simulation on FPGAs. *IEEE Transactions on Computers*, 99:1–1, 2012. DOI: 10.1109/TC.2012.121.

[351] Hang-Sheng Wang, Li-Shiuan Peh, and Sharad Malik. Power-driven design of router microarchitectures in on-chip networks. In *Proc. of the 36th International Symposium on Microarchitecture*, pages 105–116, November 2003. DOI: 10.1109/micro.2003.1253187.

[352] Hang-Sheng Wang, Xinping Zhu, Li-Shiuan Peh, and Sharad Malik. Orion: A power-performance simulator for interconnection networks. In *Proc. of the 35th International Symposium on Microarchitecture*, pages 294–305, November 2002. DOI: 10.1109/micro.2002.1176258.

[353] L. Wang, P. Kumar, K.H. Yum, and E.J. Kim. APCR: An adaptive physical channel regulator for on-chip interconnects. In *Proc. of the International Conference on Parallel Architecture and Compilation Techniques*, 2012. DOI: 10.1145/2370816.2370830.

[354] Lei Wang, Yuho Jin, Hyungjun Kim, and Eun Jung Kim. Recursive partitioning multicast: A bandwidth-efficient routing for on-chip networks. In *Proc. of the International Symposium on Networks-on-Chip*, May 2009. DOI: 10.1109/nocs.2009.5071446.

[355] Ruisheng Wang, Lizhong Chen, and Timothy Mark Pinkston. Bubble coloring: Avoiding routing- and protocol-induced deadlocks with minimal virtual channel requirement. In *Proc. of the 27th International ACM Conference on International Conference on Supercomputing*, ICS '13, pages 193–202, New York, NY, USA, 2013. ACM. DOI: 10.1145/2464996.2465436.

[356] David Wentzlaff, Patrick Griffin, Henry Hoffman, Liewei Bao, Bruce Edwards, Carl Ramey, Matthew Mattina, Chyi-Chang Miao, John Brown III, and Anant Agarwal. On-chip interconnection architecture of the Tile processor. *IEEE Micro*, 27(5):15–31, 2007. DOI: 10.1109/mm.2007.4378780.

[357] Daniel Wiklund and Dake Lui. SoCBus: Switched network on chip for hard real time embedded systems. In *Proc. of the International Parallel and Distributed Processing Symposium*, pages 8–16, April 2003. DOI: 10.1109/ipdps.2003.1213180.

[358] Wishbone. http://opencores.org/

[359] P. Wolkotte, P. Holzenspies, and G. Smit. Fast, accurate and detailed NoC simulations. In *International Symposium on Networks on Chip*, 2007. DOI: 10.1109/nocs.2007.18.

[360] Pascal T. Wolkotte, Gerard J .M Smit, Gerard K. Rauwerda, and Lodewijk T. Smit. An energy-efficient reconfigurable circuit-switched network-on-chip. In *Proc. of the 19th International Parallel and Distributed Processing Symposium*, pages 155–162, 2005. DOI: 10.1109/ipdps.2005.95.

[361] Jae-Yeon Won, Xi Chen, Paul V. Gratz, Jiang Hu, and Vassos Soteriou. Up by their bootstraps: Online learning in artificial neural networks for CMP uncore power management. In *International Symposium on High Performance Computer Architecture*, 2014. DOI: 10.1109/hpca.2014.6835941.

[362] Steven C. Woo, Moriyoshi Ohara, Evan Torrie, Jaswinder Pal Singh, and Anoop Gupta. The SPLASH-2 programs: Characterization and methodological considerations. In *Proc. of the International Symposium on Computer Architecture*, pages 24–36, June 1995. DOI: 10.1109/isca.1995.524546.

[363] Frédéric Worm, Paolo Ienne, Patrick Thiran, and Giovanni De Micheli. An adaptive low-power transmission scheme for on-chip networks. In *ISSS '02: Proceedings of the 15th international symposium on System Synthesis*, pages 92–100, New York, NY, USA, 2002. ACM. DOI: 10.1109/isss.2002.1227158.

[364] L. Wu, A. Lottarini, T. Paine, M. A. Kim, and K. A. Ross. Q100: The architecture and design of a database processing unit. In *International Conference on Architectural Support for Programming Languages and Operating Systems*, 2014. DOI: 10.1145/2541940.2541961.

[365] Yuan Xie, Jason Cong, and Sachin Sapatnekar. *Three-dimensional IC: Design, CAD, and Architecture*. Springer, 2009. DOI: 10.1007/978-1-4419-0784-4.

[366] Yi Xu, Yu Du, Bo Zhao, Xiuyi Zhou, Youtao Zhang, and Jun Yang. A low-radix and low-diameter 3D interconnection network design. In *International Symposium on High Performance Computer Architecture*, pages 30–42, February 2009. DOI: 10.1109/hpca.2009.4798234.

[367] Yi Xu, Bo Zhao, Youtao Zhang, and Jun Yang. Simple virtual channel allocation for high throughput and high frequency on-chip routers. In *Proc. of the International Symposium on High Performance Computer Architecture*, 2010. DOI: 10.1109/hpca.2010.5416640.

[368] Haofan Yang, Jyoti Tripathi, Natalie Enright Jerger, and Dan Gibson. Dodec: Random-link, low-radix on-chip networks. In *Proc. of the International Symposium on Microarchitecture*, 2014. DOI: 10.1109/micro.2014.19.

[369] Yuan Yao and Zhonghai Lu. DVFS for NoCs in CMPs: A thread voting approach. In *2016 IEEE International Symposium on High Performance Computer Architecture (HPCA)*, pages 309–320. IEEE, 2016. DOI: 10.1109/hpca.2016.7446074.

[370] Yuan Yao and Zhonghai Lu. Memory-access aware DVFS for network-on-chip in CMPs. In *2016 Design, Automation & Test in Europe Conference & Exhibition (DATE)*, pages 1433–1436. IEEE, 2016. DOI: 10.3850/9783981537079_0455.

[371] Jieming Yin, Onur Kayiran, Matthew Poremba, Gabriel Loh, and Natalie Enright Jerger. Efficient synthetic traffic models for large complex SoCs. In *Proc. of the International Symposium on High Performance Computer Architecture*, 2016. DOI: 10.1109/hpca.2016.7446073.

[372] Jieming Yin, Pingqiang Zhou, Sachin S Sapatnekar, and Antonia Zhai. Energy-efficient time-division multiplexed hybrid-switched NoC for heterogeneous multicore systems. In *Parallel and Distributed Processing Symposium, 2014 IEEE 28th International*, pages 293–303. IEEE, 2014. DOI: 10.1109/ipdps.2014.40.

[373] Kunzhi Yu, Cheng Li, Hao Li, Alex Titriku, Ayman Shafik, Binhao Wang, Zhongkai Wang, Rui Bai, Chin-Hui Chen, Marco Fiorentino, Patrick Yin Chiang, and Samuel Palermo. A 25 Gb/s hybrid-integrated silicon photonic source-synchronous receiver with microring wavelength stabilization. *IEEE Journal of Solid State Circuits*, 51(9):2129–2141, September 2016. DOI: 10.1109/jssc.2016.2582858.

[374] Bilal Zafar, Timothy Mark Pinkston, Aurelio Bermúdez, and José Duato. Deadlock-free dynamic reconfiguration over Infiniband™ networks. *Parallel Algorithms Applications*, 19(2–3):127–143, 2004. DOI: 10.1080/10637190410001725463.

[375] Lihang Zhao, Woojin Choi, Lizhong Chen, and Jeff Draper. In-network traffic regulation for transactional memory. In *International Symposium on High Performance Computer Architecture*, 2013. DOI: 10.1109/hpca.2013.6522346.

[376] Zhiping Zhou, Bing Yin, and Jurgen Michel. On-chip light sources for silicon photonics. *Light: Science & Applications*, 4:e358–, 11 2015. DOI: 10.1038/lsa.2015.131.

[377] A. K. Ziabari, J. L. Abellan, R. Ubal Tena, C. Chen, A. Joshi, and D. Kaeli. Leveraging silicon-photonic NoC for designing scalable GPUs. In *International Conference on Supercomputing*, 2015. DOI: 10.1145/2751205.2751229.

[378] Amir Kavyan Ziabari, Jose L. Abellan, Yenai Ma, Ajay Joshi, and David Kaeli. Asymmetric NoC architectures for GPU systems. In *Proc. of the International Symposium on Networks on Chip*, 2015. DOI: 10.1145/2786572.2786596.

[379] Arslan Zulfiqar, Pranay Koka, Herb Schwetman, Mikko Lipasti, Xuezhe Zheng, and Ashok V. Krishnamoorthy. Wavelength stealing: An opportunistic approach to channel sharing in multi-chip photonic interconnects. In *International Symposium on Microarchitecture*, 2013. DOI: 10.1145/2540708.2540728.

Authors' Biographies

NATALIE ENRIGHT JERGER

Natalie Enright Jerger is an Associate Professor and the Percy Edward Hart Professor of Electrical and Computer Engineering in the Edward S. Rogers Sr. Department of Electrical and Computer Engineering at the University of Toronto. She completed her Ph.D. at the University of Wisconsin-Madison in 2008. She received her Master of Science degree from the University of Wisconsin-Madison and Bachelor of Science in Computer Engineering from Purdue University in 2004 and 2002, respectively. Her research interests include multi- and many-core architectures, on-chip networks, cache coherence protocols, memory systems, and approximate computing. Her research is supported by NSERC, Intel, CFI, AMD, and Qualcomm. She was awarded an Alfred P. Sloan Research Fellowship in 2015, Borg Early Career Award in 2015, MICRO Hall of Fame in 2015, the Ontario Professional Engineers Young Engineer Medal in 2014, and the Ontario Ministry of Research and Innovation Early Researcher Award in 2012.

TUSHAR KRISHNA

Tushar Krishna is an Assistant Professor in the School of Electrical and Computer Engineering at the Georgia Institute of Technology. He received a Ph.D. in Electrical Engineering and Computer Science from Massachusetts Institute of Technology in 2014. Prior to that he received a M.S.E in Electrical Engineering from Princeton University in 2009, and a B.Tech in Electrical Engineering from the Indian Institute of Technology (IIT) Delhi in 2007. Before joining Georgia Tech in 2015, he worked as a researcher in the VSSAD Group at Intel, Massachusetts. His research interests span computer architecture, on-chip networks, heterogeneous SoCs, deep learning accelerators, and cloud networks.

LI-SHIUAN PEH

Li-Shiuan Peh is Provost's Chair Professor in the Department of Computer Science of the National University of Singapore, with a courtesy appointment in the Department of Electrical and Computer Engineering since September 2016. Previously, she was Professor of Electrical Engineering and Computer Science at MIT and was on the faculty of MIT since 2009. She was also the Associate Director for Outreach of the Singapore-MIT Alliance of Research & Technology (SMART). Prior to MIT, she was on the faculty of Princeton University from 2002. She graduated with a Ph.D. in Computer Science from Stanford University in 2001, and a B.S. in Computer Science from the National University of Singapore in 1995. Her research focuses on networked computing, in many-core chips as well as mobile wireless systems. She received the IEEE Fellow in 2017, NRF Returning Singaporean Scientist Award in 2016, ACM Distinguished Scientist Award in 2011, MICRO Hall of Fame in 2011, CRA Anita Borg Early Career Award in 2007, Sloan Research Fellowship in 2006, and the NSF CAREER award in 2003.

Printed in the United States
by Baker & Taylor Publisher Services